Earth-sheltered architecture is one answer to the energy crisis. When well-designed buildings are buried either partially or totally underground, dramatic energy savings result. Underground libraries and offices across America have trimmed their energy needs by more than 50 percent (one earth-sheltered bookstore in Minneapolis doesn't require any heat until the temperature dips below zero!). And, industries that relocate in abandoned caves discover they simply don't need any heating or cooling in their subgrade locations.

David Martindale, who has carefully researched the phenomenon of earth-sheltered architecture, offers a complete look at this promising development. Beginning with our cave-dwelling ancestors, he shows how man has utilized underground space in places such as Turkey, China, Tunisia, and the American Southwest. He discusses the uniquely American legacy of sod homes and basements and explains why many people have a psychological block about living and working underground.

Learn about the modern-day earth-shelter movement here in the United States: how it got started, who pioneered it, and why it is growing so rapidly. Spotlighting a wide variety of earth-sheltered buildings from coast to coast, Martindale shows what types of earth-sheltered structures are being built—and what savings result! He explains why underground buildings are ideally suited to tap the potential of solar energy and talks with earth-shelter home-owners—subsurface zealots who have discovered that their underground homes not only save energy, but also offer quiet, maintenance-free, environmentally harmonious living.

This comprehensive study concludes with a look into the future—a preview of tomorrow's earth-sheltered buildings and a long-range assessment of the earth-shelter phenomenon and its likely impact on the future of architecture in the United States. Illustrated with dozens of photographs, *Earth Shelters* offers a message as timely as tomorrow's headlines.

EARTH
SHELTERS

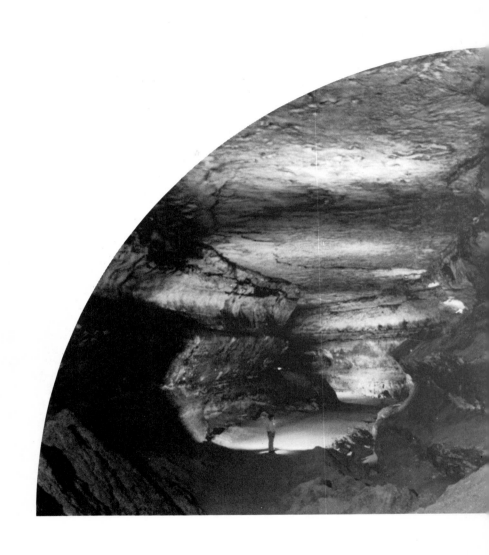

David Martindale

EARTH
SHELTERS

E. P. DUTTON · New York

690.86

Library of Congress Cataloging in Publication Data • Martindale, David. • Earth shelters. • Bibliography: p. 135 • 1. Earth-sheltered houses. I. Title. • TH 4819.E27M37 1981 728 80-22553 •

ISBN: 0-525-93199-6 (cl)   •   0-525-93200-3 (pa)

Published simultaneously in Canada by Clarke, Irwin & Company Limited, Toronto and Vancouver • Designed by Nicola Mazzella • 10 9 8 7 6 5 4 3 2 1 • First Edition

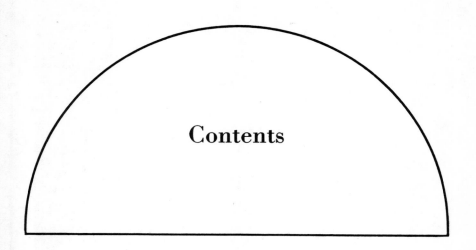

Contents

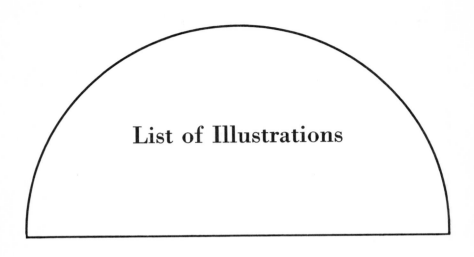

List of Illustrations

An eight-page full-color insert follows page 84.

Winston Home in New Hampshire
Clark-Nelson home, west central Wisconsin
John Barnard's Ecology House
Baldtop Dugout, New Hampshire
University of Minnesota's Williamson Hall
Atrium of the University of Minnesota's Williamson Hall
St. Benedict's Abbey
Suncave home designed by David Wright, Santa Fe, New Mexico
Kitchen of Clark-Nelson home
Davis Cave
Mark Simon's Crowell House
Seward West townhouse, Minneapolis
Two-story home in Burnsville, Minnesota
International Trade Center beneath Kansas City, Missouri
Earth-sheltered home
Forest House, central Florida

DAVID MARTINDALE is a freelance writer and photographer. He has published articles in *Reader's Digest, Psychology Today, Penthouse, Cosmopolitan, Travel & Leisure, Science Digest, Glamour,* and many others. His photographs have appeared in *Smithsonian, New Times,* and *Airline Pilot,* to name a few. Many of the photographs for this book were taken by him. At present, Mr. Martindale lives in Washington, D.C.

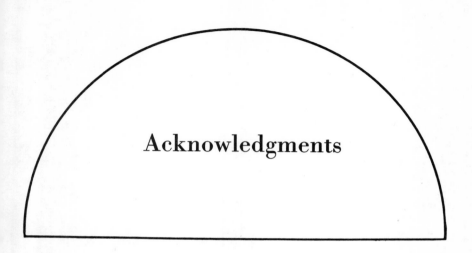

Acknowledgments

Although it's impossible to list all the people who assisted me in the preparation of this book, I would like to single out the following people for special thanks. In alphabetical order: John Barnard, Dr. Thomas Bligh, Dr. Lester Boyer, Ken Labs, Don Metz, Frank Moreland, William and Bunny Morgan, Dr. Truman Stauffer, Dr. Ray Sterling, and Malcolm Wells.

David Martindale

EARTH
SHELTERS

1
The Ancient Art
of Living Underground

cave men: n. primitive, club-wielding savages, crude in manner, vulgar in taste; last seen thirty-five thousand years ago.

Quite obviously, no dictionary describes our Stone Age ancestors in quite these unflattering terms. Yet such a definition, however inaccurate, does reflect a widely held stereotype of cave men. Yet the cranial capacity of some cave dwellers was actually larger than that of twentieth-century man, and these "savages" employed survival skills and strategies every bit as ingenious as today's corporate games.

Despite all the evidence to the contrary, the image remains: cave men (and cave women, as well) were brutal, coarse, and downright ugly. Above all, they were dumb. After all, the reasoning goes, no one would *want* to live in a cave!

Whether we live in penthouses or bungalows, stuccoes or tract homes, we tend to snub our noses at both the cave dweller and the cave itself. We ignore its origins and overlook its basic practicality. Caves are found in nearly every corner of the world and in every climate. Some caves develop as sea water beats relentlessly against cliffs, while others are spawned by lava ruptures and earthquakes. Most caves are formed by the slow, steady action of rainwater on soluble rocks such as limestone and gypsum. A combination of erosion and rain's slightly acidic content act to widen natural cracks in the rock until, eventually, a cave results.

Far from being dingy grottoes of last resort, caves proved to be prac-

Temperatures inside Mammoth Cave in Kentucky remain a near-constant 54 degrees F. year round. (Courtesy: National Park Service.)

tical habitats which boasted a variety of benefits. The temperature inside most caves is nearly constant year round; Mammoth Cave in Kentucky, for example, maintains a nearly stable fifty-four degrees Fahrenheit summer and winter. Not exactly within the ideal range of human comfort, but far less severe than the unprotected outdoors.

Caves also provide protection against natural calamities such as fires, hurricanes, tornadoes, and lightning. And in the world of the Neanderthal, caves offered not only a superb place in which to hide, but also a strategic location from which our ancestors could defend themselves when attacked.

And perhaps best of all, since nature was both the architect and the contractor, the Stone Age cave dweller never had to worry about design or building materials: no architects or blueprints were required, nor was there any need for cement, masonry, or wooden beams. All the occupants had to do was walk in, light a fire, and their cave became their castle.

Unfortunately, we don't know a great deal about these early cave dwellers. Prehistoric man, by definition, left no detailed narratives describing day-to-day life in the caves. Occasionally, the more artistic Stone Agers grabbed a flint and some pigments and sketched a mural on a cavern wall, but these illustrations said less about daily cave existence than about the four-legged mammals who were the subjects of man's first art exhibit.

Although our knowledge is limited, we do know this: Stone Age man was nomadic. Few cave dwellers called one cave home for a lifetime; they moved about the countryside, hunting for meat and collecting fruits and vegetables. When weather permitted, they lived in the open. When they gathered in caves, they probably spent most of their time near the mouth, which provided access to daylight. Here, the smoke from their fires could easily escape into the open, clearing the air and frightening away predators.

From the standpoint of basic habitation, caves made sense; what they lacked in comforts they more than compensated for in security, temperature moderation, and protection from the elements. They could be occupied at a moment's notice and were easy to maintain.

Then why did Stone Age man eventually abandon caves in favor of surface dwellings? Anthropologists aren't quite sure. Most likely, lack of mobility was the principal factor. Eventually, the hunting near caves would become limited, the food supplies scarce. Yet hunting and gathering away from the caves meant spending long periods of time unprotected. The solution was obvious: construct a shelter on the surface. And as man slowly learned to improve his skills with tools, free-standing surface structures became a more viable option.

Man never completely abandoned his love affair with the cave. On the contrary, some caves have been occupied almost continuously since prehistoric times. But once *Homo sapiens* began building their own shelters, living in caves—or beneath the soil or in rock-carved caverns—became the exception rather than the rule. This distinction between surface and subsurface inhabitants even entered our vocabulary: prehistoric cave dwellers are known as *troglodytes.*

Despite the fact that the civilizations of Egypt, Greece, and Rome were not troglodytic, the underground tradition survived and even flourished throughout different cultures. Not surprisingly, the reasons these subterranean pockets developed were often remarkably similar.

For example, Israeli archaeologists have uncovered an early troglodytic settlement in the Negev Desert. Believed to date back more than five thousand years, this community consisted of an entire underground city, complete with a maze of winding tunnels leading to large rooms and storage areas. Discoverers speculate this subgrade Negev outpost was probably built to provide a refuge from the heat and a shield from the desert sands.

The desire to modify a harsh climate often played a crucial role in the decision to live underground. Consider the Berbers who settled in southern Tunisia; confronted with the extreme heat of the blistering Sahara, these Moslem tribesmen excavated a layer of sandstone and constructed underground cities. At the subterranean town of Matmata, for example, five to six thousand Berbers once lived in such seclusion that their presence could barely be detected from the surface (a distinct military advantage which the Berbers did not overlook). Most of the dwellings at Matmata were built around a central courtyard approximately forty feet square and forty feet deep. The entrance was via tunnels burrowed into nearby hills. Rooms fanned out from the central courtyard, and although each had access to light, no room received the full direct rays of the desert sun. For this reason, these subterranean homes remained much cooler than any structure which could have been built on the Tunisian surface. The proof of their practicality is that some are still occupied today.

The Romans emulated the Berbers when they occupied Bulla Regia in northern Tunisia. Using conventional building materials, they constructed Roman peristyle homes below grade. They also designed a few underground retreats for their emperors in Italy. In these cool underground grottoes, royalty could escape the oppressive summer heat of Rome.

Submerged settlements have also shielded man from the winter cold. Cappadocia, a barren, mountainous region of the Goreme valley in central Turkey, is plagued by Arctic-style winters and summers nearly as torrid as those in Death Valley. Cappadocia is one of the least hospitable spots on the planet. Contorted by volcanic activity, wracked by erosion, the land is nearly devoid of vegetation; a few stunted trees are all that dot the horizon. Building materials of any kind are sparse, and lumber is virtually nonexistent.

The Cappadocia region of Turkey is famous for its rock-hewed dwellings, like these in Goreme. (Courtesy: Turkish Tourism and Information Office.)

Yet since the third century, the rugged villagers of this poor agricultural area have lived continuously in forty-one underground cities scattered throughout Cappadocia. For centuries, historians have speculated as to why the Cappadocians decided to burrow underground. No one knows for sure; perhaps the early Christians who first built these cities sought a safe refuge from their enemies. But while their primary purpose may have been defensive, such subterranean communities also provided a superb adaptation to a harsh environment.

By any subsurface standards, the cities of Cappadocia are huge. Some extend for more than three miles, and have sheltered as many as thirty thousand inhabitants living on ten levels. Some levels are reserved for living quarters, while others are for livestock, storage, or wineries. Vertical and horizontal tunnels criss-cross in a seemingly helter-skelter fashion, feeding into rooms and openings which vary in size and shape. Ventilation shafts rise from the bottom floors to the top. Often, amenities such as cupboards, furniture, stoves, and even box beds are made of stone, easily carved from the same porous volcanic rock which forms the walls, roofs, and floors.

Not only did the Cappadocians cultivate a centuries-old tradition, but they also inhabited what is surely some of nature's most bizarre geological creations. Throughout this region of Turkey, more than twenty thousand anthill-shaped cones of volcanic rock sprout above the landscape. Although carved through erosion, they were hollowed out and turned into apartment complexes by the Cappadocians. Indeed, some of these mounds contain as many as six levels of living quarters, each excavated with the same care as the subterranean cities. And like their underground counterparts, the interior temperatures of these cones remain nearly stable year round despite the bone-chilling, sun-scorching fury of the climate.

Because the region was an early Christian stronghold, no fewer than 365 Byzantine-style underground churches remain scattered throughout the area. And the Cappadocians weren't the only ones to bury their

Indian ruins like these in Mesa Verde National Park in southwestern Colorado provided protection as well as relief from extremes in temperature. (Courtesy: Colorado State Office of Tourism.)

churches or carve them into stone. In Ethiopia, rock-cut Coptic churches were chiseled into hillsides at spectacular locations. Others were built inside great natural caves as early as the fourth century. At Ajanta, India, more than two dozen Buddhist temples were also constructed inside caves, the oldest such place of worship dating back more than seventeen hundred years.

Was there any religious significance to the fact that such sanctums were subsurface? Perhaps; monastic cloisterers and religious zealots might have been attracted to the otherworldliness of their locations. At the same time, troglodytic churchmen might have decided to go underground to secure an isolated retreat from the everyday hustle and bustle of secular pursuits. Their precise motivations remain mysteries.

Possibly the best-known underground religious gathering places are the kivas used by the Pueblo Indians of the American Southwest. First built as early as 300 A.D., a kiva consisted of a stone or clay-walled chamber usually circular in shape. Buried several feet underground, the rooms varied from ten to eighty feet in diameter. Most kivas had an external ventilating shaft, and nearly all were covered by an earthen dome which allowed some light to enter. Although its exact purpose is unclear, the kiva is believed to have been a ceremonial meeting place for male members of the tribe.

The Pueblos had a long troglodytic tradition. Before 300 A.D., nearly all tribal dwellings were buried beneath earthen mounds. Later, the Pueblos carved their homes into cliff caves in what is now Arizona, Colorado, and New Mexico. Accessible only by ropes and wooden ladders, these locations proved as effective in providing a secure defense as they did in shielding Pueblo villagers from extremes in temperatures.

There is nothing new about human beings living and working underground; the practice predates history and has continued into modern times. And although the earth-sheltered structures we build today barely resemble the dwellings of ancient troglodytes, the reasons for going underground have hardly changed. We, too, seek protection from extremes in temperature, a refuge from intruders, and a long-lasting, solid edifice which requires little or no upkeep.

The advantages of going underground have yet to result in widespread support for the earth-shelter option. Earth-sheltered structures have an undeserved image problem: they are not yet "socially acceptable." Rationally, earth shelters can withstand the test of logic and reason; yet when pitted against the suspicion that "down is dismal," support crumbles. And negative feelings toward the underground have been reinforced by the American experience.

2
From Basements to Bomb Shelters: Underground Gets a Bad Name

It is too bad that the tunnel-digging Cappadocians weren't avid travelers. Had they abandoned Turkey, crossed the Atlantic, and burrowed beneath North America, we might have learned to appreciate underground space rather than fear it.

But the European settlers of North America were not troglodytes. And land in America was plentiful and far from barren. Building materials were as abundant as the forests. The climate was similar to that of Europe, and the settlers already had a rich tradition of surface habitats.

The early American colonists did construct at least one type of enduring underground landmark: the cellar. From Virginia to New England, the colonial countryside was dotted with these subsurface storage vaults where beets, potatoes, and other produce were preserved. Not all cellars were built alike. Root cellars were banked with earth and topped with sod, while ground cellars were actually built below grade. Both provided excellent year-round thermal protection for foods. They also doubled as storm shelters; by the time cellars appeared in the Midwest, they were often used to escape the fury of tornadoes.

Handy little areas, those cellars. But live there permanently? Forget it!

As homesteading pushed America westward, many pioneers arrived on the Great Plains owning little more than the land. Some were so poor they couldn't afford to transport building materials, much less buy any. So they improvised; using soil and grass, these pioneers built homes. Most

The sod home of J. D. Troyer near Callaway, Nebraska, 1892. (Courtesy: Solomon D. Butcher Collection, Nebraska State Historical Society.)

of these sod huts or dugouts were free-standing structures made of sod blocks two to four inches thick. Some huts were carved into the side of a hill or ravine. In such locations, they moderated the extreme temperatures of the Plains.

Of course, none of these sod homes was ever a candidate for architectural recognition. Most deteriorated rapidly and required constant repair. But few were intended to be permanent, and sooner or later, homesteaders built conventional homes on the surface. When they did, they often kept their sod house and converted it to a barn or a crop-storage area.

Enter the era of the basement.

In addition to serving the same storage function as the ground cellar, basements are designed to insulate the air space between a warm house and the cold earth. Yet they have often been a headache to their owners. If a basement doesn't leak, then it floods. And since many basements are damp, dark, and poorly ventilated, it is little wonder that they are held in such dismal regard. Of course, with dehumidifiers, improved illumination, and better air-circulation systems, basements can become every bit as livable as surface dwellings. Yet the "basement bias" remains.

There is certainly no status attached to living in a basement apartment. In fact, your mother probably warned you against renting one. High-level executives are seldom ensconced in plush basement hideaways. And no one has to tell a presidential aide that he has lost clout with the boss when he arrives to work one morning and discovers his office has been shifted to the White House basement.

Basement phobia is reflected in American building codes. Following World War II, home-buying veterans could sometimes afford to purchase a foundation, but not the home itself. Some built a basement and moved in, determined to top off the structure in the near future. Still strapped for cash years later, they found that their basement shells had become near-permanent fixtures. In many communities, health officials convinced zoning commissioners that such long-term subsurface habitation was at best weird and at worst a health hazard. As a result, living in a basement shell remains unlawful in many communities.

Even architect Frank Lloyd Wright shared the notion that basements and good health simply weren't compatible. "A house should—ordinarily—not have a basement," wrote Wright in *The Natural House.* "In spite of everything you do, a basement is a noisome, gaseous damp place. From it came damp atmospheres and unhealthful conditions. . . ."

Cellars, dugouts, and basements do not provide a very positive impression of underground space, though they're hardly sufficient to account for today's antisubterranean bias. What is it about being underground which conjures up such negative emotions? Why do people of diverse backgrounds reject living and working beneath the earth?

Some of this apprehensiveness may be linked to the fear of death. When we die, what happens to our bodies? More often than not, they are buried (a word with negative connotations if ever there was one). Bodies are then interred beneath the surface and covered with earth. Small wonder that earth-sheltered buildings remind some people of catacombs.

Besides being linked with death, the underground also has profound implications for those who believe in a hereafter. The Bible is quite clear about the location of heaven. According to Genesis, when Jacob dreamed, he beheld "a ladder set up in the earth, and the top of it reached to heaven. . . ." And just as clearly, the Scriptures tell us where to locate hell. The fact that the torments of fire and brimstone are conducted below ground is not lost on the young and impressionable, nor does it necessarily vanish with adulthood.

We need not enroll in a seminary to fully appreciate the concept of hell on earth. And that earthly hell is frequently located underground; ask any subway commuter in New York. All but the most modern subways suffer from negative reputations. So do sewers and mines.

Perhaps the problem stems from the absence of positive underground experiences. If couples wed underground, if gourmets dined underground, if musicians composed underground, would we still wrinkle our brows at the thought of living and breathing below the surface?

To what extent does our language contribute to this link between the underground and the inferior? You say you suffered an embarrassing *put down* when the boss caught you in a *cover up?* Well, don't let it get you

If all subway systems were as clean and modern as Washington, D.C.'s Metro, commuting underground would be a pleasant experience. (Photo: David Martindale.)

down. It won't mean your corporate *downfall.* Just be thankful you didn't *sink* so *low* that you joined the *underworld.* But then, that would have been *beneath* you.

Certainly much of the underground's negative stereotypes can be traced to military applications of subterranean space. More than a hundred years ago, military planners began constructing "bombproofs," underground shelters designed to protect troops against enemy artillery and mortars.

Not long afterward, man learned to fly. But with that discovery came a new horror: aerial bombs to terrorize soldiers and civilians alike. Troops in World War I often burrowed deep underground for protection, and within a generation, whole cities were fleeing aerial holocausts by scurrying below the surface. Who can forget the newsreels of Londoners marching quickly but somberly into bomb shelters and subway stations to escape the fiery nightmare of German V-2s?

The military uses of underground space were not confined to the West. Commanders in China, Japan, and Korea developed elaborate networks of underground bunkers, sometimes constructing entire self-contained cities beneath the ground. Subterranean networks played a key role during the Vietnam War. Excavated by the Viet Cong and North Vietnamese in order to escape daily air strikes, these tunnels were nearly impossible to detect or destroy until the United States began dropping "earthquake" bombs which shattered enemy burrows by concussion.

At no point in recent history did the underground option become more chilling than during the late 1950s and early 1960s: the era of the fallout shelter. Saber-rattling by the world superpowers and the resulting heightening of cold-war tensions suddenly raised the ominous specter of an all-out thermonuclear confrontation. And for a time, it seemed as though the survival of the United States depended upon our ability to tap effectively our underground resources (an assumption which might still be valid today).

At last Americans were being schooled in the value of underground space. But what a schooling!

Civil-defense authorities were quick to point out that adequate underground shelter provides nearly complete protection from radioactive fallout and thermal radiation. It also provides a good deal of protection against blast waves. But although the fallout shelter promised to protect, the link between underground space and swift annihilation further prejudiced a people who never were enthusiastic over going underground in the first place.

As the cold war intensified, the United States geared up for a mas-

Built in 1960, this 150-person fallout shelter in Thomasville, Georgia is covered with two feet of earth. (Courtesy: Hoesch America.)

sive civil-defense program. Neutral Sweden had already done the same thing. Beneath their soil, the Swedes had carved underground hangars for jet airplanes, massive rock-hewn sea pens for fiord-based destroyers, and huge granite shields to protect repair shops, barracks, and fuel dumps. In addition, the Swedes constructed elaborate underground fallout shelters, some capable of holding as many as twenty thousand people. During peacetime, these great subterranean chambers have been used for a variety of commercial and industrial pursuits (see page 108).

Despite its perception of the threat, the United States did not match the scale of the Swedish effort. But many in government did feel that civil defense deserved a much greater priority than it had been receiving. As President Kennedy said following the 1961 Berlin crisis, "To recognize the possibilities of nuclear war in the missile age without our citizens knowing what they should do and where they should go would be a failure of responsibility."

So the bureaucracy prepared to fill this information gap. Pamphlets were printed, spokespeople were briefed, studies were commissioned. Not only did Washington encourage state and local governments to build their own shelters, but individual Americans were also urged to do the same thing. "Civil defense must be a part of the normal way of life," government CD official Virgil Couch told *Time* magazine. "Like smallpox vaccinations, we've got to get used to it and build it into the normal fabric of our lives. In the old days, for instance, outhouses sat in the backyard; now the bathroom has moved inside. Garages used to sit on the edge of the lot; now many garages have been built into the home. The next room to follow this pattern is the family fallout shelter."

And so the Great Dig began. Manufacturing firms labored overtime to crank out elaborate shelters, as well as more makeshift models, some of which sold for as little as $150. Almost overnight, "survival stores" opened to tap a boom business selling shelter supplies. These were accompanied by the inevitable doomsday hucksters, hawking such useless remedies as "fallout salves" and "antiradiation pills."

Despite all the hype and hysteria, fallout shelters were a flop. Americans simply refused to make the bomb shelter as ubiquitous as the backyard swimming pool. A number of earth-sheltered structures were built during this period, most very unimaginatively designed. By the time the fuss died down, it seemed the only fallout from the whole affair was yet another black eye for underground space.

Interestingly enough, the concept of the personal fallout shelter enjoyed a brief revival a few years later at the 1964–1965 New York World's Fair. The man behind the revival was Jay Swayze, a Plainview, Texas home builder. Deeply concerned about the threat of nuclear confrontation, Swayze came up with a simple idea: why not create a truly livable bomb shelter, one which could double as a conventional residence year around? Eventually this idea was transformed into a reality, first as Swayze's home in Plainview, and later as an exhibit at the world's fair.

Entered from a spiral staircase atop a roof garden, the exhibit home consisted of a waterproof, radiationproof concrete shell. Inside the shell was a conventional style ten-room home surrounded by a narrow corridor. This entire ship-in-a-bottle design was buried three feet underground. Colorful murals of seascapes, mountains, forests, and meadows were painted on the outer vault walls, and a special lighting system was designed to duplicate the various color changes of natural sunlight. "Outside" the home, flowers bloomed beside a splashing fountain, the focal point of Swayze's underground backyard.

No one can say whether Jay Swayze advanced or hindered the cause of the earth-shelter movement. His exhibit did manage to raise a few eyebrows at the fair. Consider the windows in his underground home: they were real enough, opening and closing like conventional windows. Yet they lacked one important feature—an outdoor view.

To Swayze, the benefits of protection from nuclear fallout clearly outweighed the need for direct sunlight and a view. But not everyone would have the same reaction to living in a viewless concrete vault. Some occupants would probably feel a bit uneasy and others would experience a great deal of anxiety, even claustrophobia. The fact remains that no one knows for sure what reactions are most likely; although studies have been done on emotional and sensory deprivation, no one has investigated the long-range psychological and physical effects—if any—resulting from living permanently underground in a Swayze-type environment.

As for windows, there seem to be two reasons why we hold them in such high esteem. Firstly, windows allow daylight to enter, and daylight seems to brighten our spirits and help us keep our body's daily rhythms in sync. Secondly, windows give us something to look at. They increase the flow of information to our brains, thus heightening our level of stimulation. And as a 1975 National Bureau of Standards study observed, "Although a view out is generally desirable, in some restricted and monotonous situations, it becomes almost a necessity."

Windows also have disadvantages. Improperly positioned, they contribute to excessive heat gain which results in wasted energy. Windows produce glare. They allow sunlight to wake us in the morning, and they often prevent us from maintaining our privacy. Windows break and require constant cleaning. And windows also have the nasty habit of providing us with views of belching steel mills rather than of pristine mountains.

There seems little doubt that we are a prowindow society. Corporate clout can practically be measured in panes of glass. Office pecking orders are determined by whose desk sits beside how large a window, even though high-rise windows serve no utilitarian function: lighting is fluorescent and ventilation is pumped air. Windows are status symbols, and seem to fill a pressing human need linked to vision.

Had Jay Swayze recognized this prowindow bias and incorporated it into his underground homes, perhaps he would have become the father of the earth-shelter movement. Instead, his idea waned, although he remains among its most adamant proponents.

The legacy of ventures like Swayze's lingers today: in the public mind, going underground is usually synonymous with a windowless environment. In most cases, nothing could be further from the truth. But even when that assumption *is* correct, it misses a key point.

In the early 1970s, mechanical engineer and earth-shelter advocate Dr. Thomas Bligh asked students at the University of Minnesota to poll eight hundred office workers in a St. Paul building. The question: would you like to work in an underground building? Not surprisingly, all eight hundred said no. Only after it was pointed out that they already worked in a windowless building did many change their minds and admit that going underground might not be a bad idea. Considering the fact that such surface buildings as shopping centers, theaters, restaurants, museums, factories, and laboratories are frequently devoid of windows, it is difficult to fathom why a similar environment below ground would make any difference. The bottom line to such a building's acceptance has less to do with its location than with how tastefully and imaginatively its interior is designed.

The architects and owners of the first modern earth-sheltered buildings were well aware of the underground's nasty reputation. They had heard all the snide comments, and knew their friends and neighbors considered them misguided twentieth-century troglodytes.

These modern architectural pioneers realized that if the earth-shelter option were ever going to gain widespread acceptance, their buildings would have to dispell the myths. Not an easy task, to be sure. But in the end, they created bright, cheery, sun-bathed structures which, from the inside, appeared every bit as conventional as their surface counterparts.

After centuries of abuse, the underground was vindicated.

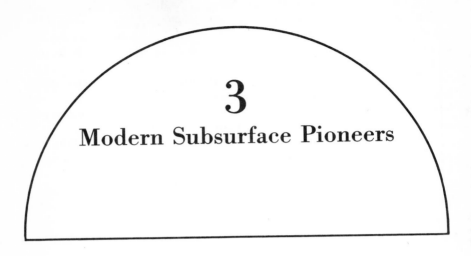

3
Modern Subsurface Pioneers

Long before Jay Swayze burrowed his way into the hearts and minds of New York World's Fair goers, a little-known, tenacious Italian immigrant carved an elaborate earth-sheltered labyrinth which assures him permanent mention in an underground hall of fame.

His name was Baldasare Forestiere. Beginning in 1908, this maverick tunneler spent thirty-eight years carefully excavating the rock beneath the barren desert soil of Fresno, California. He was still digging when he died in 1946.

What Forestiere left behind was a sixty-five-room maze of catacomb-like chambers which stretched beneath seven acres of Fresno. Roman-style courtyards pierced the surface, bathing surrounding rooms in sunlight. Elsewhere, glass skylights offered additional contact with the outside world. Vaulted arches shaped long, winding passageways, and lush plants and vines covered grotto walls.

No one knows why Forestiere chose to chisel such an ambitious engineering and architectural marvel. Although some rooms served as living quarters, the scale of excavation was out of proportion to need. Certainly Fresno's harsh climate was one factor which prompted Forestiere to go underground. Built ten to thirty-five feet beneath the surface, the home deftly managed to escape the scorching 120-degree heat on the surface. Even at the height of the summer, temperatures in Forestiere's grottoes seldom exceeded 70 degrees. Fireplaces provided what little warmth was needed in winter.

Baldasare Forestiere's desire to escape the relentless Fresno sun was not motivated by steep utility bills or a desire to conserve energy. Instead, digging underground was just something which made sense—an innovative but remarkably practical solution to a basic problem of human comfort. Yet instead of viewing Forestiere as a visionary, a man with an idea that worked, most people looked on this earth-shelter pioneer as a peculiar iconoclast with a morbid fascination for cowering beneath the ground. Forestiere died in obscurity, gaining nary a convert to the earth-shelter cause.

Not all early dabblers in earth-sheltered architecture were as unheralded as Baldasare Forestiere. For example, throughout his long career, Frank Lloyd Wright maintained a deep respect for the sanctity of the earth. Rather than the land being subordinated to the building, Wright believed the building should be shaped to its terrain. And while nearly all of Wright's creations rested atop the terrain, at least two were bona fide earth-sheltered structures.

In 1942, Wright built the second Herbert Jacobs home in Middleton, Wisconsin. From the north, this curved, two-story residence looks like nothing more than a hill penetrated by a stone doorway. But inside, the south-facing windows bathe the living area in sunshine and provide a sweeping view of the countryside. Bedrooms are located on an overhanging balcony. Wright employed this same into-the-earth concept when he designed the Cabaret Theater at Taliesen West near Phoenix, Arizona. Built in 1950, the building is a striking example of wood and stone nestled partially underground to provide a subdued desert profile unattainable with a conventional surface structure.

Despite their original designs, the Jacobs home and Taliesen West proved little more than peripheral in a career renowned for striking surface dwellings. Until Jay Swayze pioneered the modern underground home in 1964, the earth-shelter option remained largely ignored.

Between 1964 and the late 1970s, when the underground movement blossomed, earth-sheltered architecture received a pre–energy-crunch boost from five creative architects. Their backgrounds varied, as did their reasons for going underground. Although they each recognized that greater energy efficiency was a by-product of their unusual designs, most were attracted to the underground option by advantages that had little to do with energy—advantages which only add to the appeal now that fuel is increasingly difficult to obtain and expensive.

MALCOLM WELLS

When Malcolm Wells began his career as a civil engineer, he never dreamed that he would become the leading proponent of underground buildings; he never even anticipated becoming an architect. Although he lacked formal architectural training, Wells worked closely with architects

Frank Lloyd Wright's Cabaret Theater at Taliesen West in Scottsdale, Arizona. (Courtesy: Yale Art and Architecture Library Photograph Collection.)

when his career began. Eventually, he began designing buildings of his own, and in time, he became a highly successful conventional architect who produced what he now calls "junk" for major corporate clients.

For Malcolm Wells, the first step on his road to underground revival was a 1959 visit to Taliesen West. Wells was impressed not only by the unique design of Wright's Arizona hideaway, but also by the fact that the building remained cool despite the desert heart. Five years passed. Wells continued designing surface buildings, including the RCA pavilion at the 1964 New York World's Fair. Wells had heard about Jay Swayze's underground home: about the artificial sunlight, the mechanical breezes, the windows which opened onto painted forests. Convinced such a home had little appeal, Wells did not even visit Swayze's exhibit.

It wasn't until later in 1964 that Malcolm Wells formally converted to the underground cause. In what was to be the first of many such articles, the Cherry Hill, New Jersey architect wrote a now-famous piece called "Nowhere to Go But Down" for the February 1965 issue of *Progressive Architecture.* In it, he argued that what architects did to the land was every bit as destructive as what hunters had done to the American buffalo. No home, said Wells, was as beautiful as a forest. The only true reality was nature. And in order to preserve nature, Wells concluded, man would simply have to build underground.

To Wells, the advantages of building beneath the soil were obvious. Earth-sheltered structures protect the land by preventing erosion and allowing rainwater to percolate into the earth. They produce greenery and encourage wildlife habitation. Underground buildings are quiet, offering protection from such aural intrusions as street traffic and barking dogs. They also require very little outdoor maintenance. And as Wells was

quick to point out, underground buildings allow greater density without a corresponding feeling of crowdedness.

Malcolm Wells became so convinced of the rightness of his cause that he stopped designing conventional buildings. Since these were the only kind his clients preferred, Wells soon found himself without any clients. Like a missionary who accepts adversity as a badge of honor, Wells remained undaunted. Instead of designing buildings, he became an architectural proselytizer for the underground cause. He wrote articles, lectured, and drafted sketches of futuristic underground structures. Although he gained considerable notoriety as an architectural rebel, his early pronouncements met with skepticism and outright disbelief. Like Baldasare Forestiere, Malcolm Wells was ahead of his time.

Six years passed before Wells translated his earth-shelter concepts into reality. In 1970, he constructed an underground office next to his home in Cherry Hill. Besides offering a wealth of practical data, the earth-sheltered office provided Wells with a far greater appreciation of the energy-saving aspects of underground buildings. In 1975, Wells used this knowledge when he designed a 2,400-square-foot solar heated home called Solaria in nearby Indian Mills, New Jersey, a home the architect described as a "hill covered with wildflowers." When he moved to Cape Cod in 1979, Wells constructed an elaborate underground home and office with overhead skylights in Brewster, Massachusetts, intended as a showcase of innovative design.

Thanks to projects such as these, as well as to the increasing acceptance of the underground concept which he helped pioneer, Malcolm Wells is no longer without clients. He continues to write and lecture, but most of his time is spent consulting and designing underground projects.

Nature has reclaimed the roof of Malcolm Wells's home, Solaria, in Indian Mills, New Jersey. (Photo: Bob Homan.)

His once unpopular idea has been vindicated; the prophet of underground living is now one of its chief purveyors.

PHILIP JOHNSON

Unlike Malcolm Wells, Philip Johnson is not an architectural maverick. Ever since the 1930s, Johnson has ranked as the dean of the international style of modern architecture, designing some of America's most famous structures, including the Seagram's building in New York, Penzoil Place in Houston, and the IDS Center in Minneapolis. Johnson is the epitome of orthodox American architecture.

Yet during the course of his long career, Philip Johnson decided to bury two of his creations underground. The most famous is the Geier house in suburban Cincinnati.

Named for its owners, this 4,000-square-foot home was built in 1965.

It sits just six inches above a man-made lake and is covered by fifteen inches of earth. A narrow entranceway is all that is visible from the parking pad in front of the Geier home. Poking from atop the earthen roof are several rust-colored cylinders, each of which hides such mundane necessities as exhausts, chimneys, and skylights.

Despite its earth cover, Philip Johnson doesn't consider the Geier house a bona fide underground building. The reason: instead of using the earth as a form of protection, Johnson used it as a shaping device. "It was a beautiful field," Johnson told the *AIA Journal* in 1978, "and I felt its connection with the water was so important that why not keep it just the way it was. . . . The dwelling affords complete privacy from the main road, and with the water coming to the house as it does, you have the feeling of being on an island away from the world, although you're in the middle of suburbia. All that the Geiers can see looking out their windows is their own land rising on the other bank."

A year after the Geier house was completed, Johnson made a similar underground foray when he unveiled his earth-sheltered art museum. Located at his home in New Canaan, Connecticut, the museum was built underground for two reasons. The first was practical: because it afforded protection from the elements, the museum was able to achieve maximum environmental control. The second reason was a bit more romantic: Johnson didn't want his backyard dominated by a building. For this reason, he molded a gently rising mound penetrated by a single small door.

Although he designed two of the earliest earth-sheltered structures in America, Philip Johnson never considered himself an earth-shelter architect. Nor did he push for the construction of more underground buildings. Yet because of his considerable prestige, Philip Johnson drew some attention to a budding new architectural movement at a time when most critics merely scoffed at the idea.

DON METZ

In the early 1970s, Don Metz decided he wanted to build a speculative home on property he owned in New Hampshire's White Mountains. The site he selected was midway up a south-facing hill which commanded a spectacular fifty-mile vista of the Connecticut River valley. Because the site was so breathtaking, the young Yale architectural graduate was determined not to mar the hill with an ugly structure which would stand out rather than blend in. For this reason, he opted to build underground.

"I didn't want a big architectural statement imposed upon the land," recalls Metz. "The underground configuration on that site really made the most sense." It also made sense to its prospective owners, Mr. and Mrs. Oliver Winston, who purchased the home before it was completed in 1972.

Sliced into the side of the hill, the Winston house is a 2,800-square-

Designed by Philip John-son, the earth-sheltered Geier house rests just above a man-made lake. (Photo: Ezra Stoller.)

foot rectangle with a sixty-foot south-facing wall dominated by glass. Because the bedrooms and the living area are positioned along this southerly exposure, they are able to take advantage of both the view and the solar energy which streams through the windows in the winter. During the summer, carefully designed overhangs prevent the sweltering summer rays from penetrating the building. Utility and service areas are nestled toward the back of the home, and a dining room sits squarely beneath a skylight.

As Metz intended, the Winston house became an integral part of the New Hampshire landscape. Looking down on the home from farther up the hill, wildflowers and grasses have reclaimed the fifteen inches of earth cover on the roof. Only the skylight and service ducts which poke unobtrusively in the air give any hint of human habitation below.

The Winston home proved an ambitious project for Don Metz. Not only did he design the house, but he also supervised most of the construction. The effort paid off: as a tribute to its novel design, the Winston home was voted an Architectural Record House of 1974. The structure received so much favorable public attention that Metz began designing other earth-sheltered homes. By designing and building a quality earth-sheltered home like the Winston house, Don Metz helped to dispel the notion that underground buildings are cold, dark, and forbidding structures. He created a home that rivals any surface dwelling for brightness, warmth, and downright livability.

MICHAEL McGUIRE

A successful Stillwater, Minnesota architect, Michael McGuire has won several awards for architectural excellence. He has also earned considerable local acclaim. Despite the fact that nearly all of his designs are for surface structures, Michael McGuire is probably most famous for a single earth-sheltered creation.

It is called the Clark-Nelson home, named for the two professors who commissioned McGuire to design this experimental home. By any measure, it is a most unusual residence.

Set into a hillside near the St. Croix River in western Wisconsin, the home is dominated by two parallel steel culverts. The smaller tube is thirty-eight feet long and forms the living and dining areas as well as the kitchen. The longer fifty-eight-foot tube contains two complete apartments, each with separate living and sleeping quarters. Laundry and mechanical systems are located in a connecting culvert which links the two tubes.

The striking futuristic appearance of the Clark-Nelson home is not confined to the exterior; textured polyurethane walls undulate throughout the home, avoiding straight lines, flat surfaces, and harsh rectangles. As for light, it not only floods the home at each end of both tubes, but it also pierces the six-inch earth cover, penetrating the home via several strategically placed skylights.

When McGuire set out to design the home, he wasn't sure what type of structure would result. However, he was determined to minimize the impact of the home so that he could preserve the natural setting. Then one day, while on a picnic in a nearby state park, he noticed a steel culvert being used as a road bed to ford a stream. Suddenly, McGuire had an idea. The arching effect of the steel culverts could support large earth loads. So why not put them to work in an underground residence?

Today, the culverts are nearly invisible. The land surrounding the Clark-Nelson home is increasingly being reclaimed by a thick growth of plants and foliage. McGuire's design has succeeded in preserving the natural setting. It also saves energy, trimming heating and cooling needs by a hefty 25 percent.

Ever since the Clark-Nelson home was built in 1972, the house has attracted considerable local, national, and international attention. McGuire told *Earth Shelter Digest* that the number of visitors to the home has been "numbing." Considering McGuire's bold design, it is little wonder the Clark-Nelson home has become not only a household word among underground buffs, but also a classic yardstick by which other subterranean homes are measured.

JOHN BARNARD

When John Barnard was a boy, he knew he wanted to become an architect, like his father, and often accompanied his father on various assignments. One day, the senior Barnard told young John that if he were to dig deep enough, he would find that the temperature of the earth maintains a near constant sixty degrees. John Barnard was fascinated. "Then why don't people build homes underground?" he asked. "It's simply not done that way," said his father, quickly dismissing the idea as ridiculous.

Even after John Barnard graduated from architectural school and began a business of his own, the idea of building underground homes continued to intrigue him. It made so much sense, he thought. Why should surface buildings freeze during winters and roast in the summers? There was such a logical alternative. Yet for years, Barnard was unable to settle on a suitable underground design.

The idea jelled while Barnard and his wife, Barbara, were on a trip to the ruins of Pompeii. Stopping at an old Roman atrium-style café, Barnard suddenly realized that a home built underground around an atrium could not only take advantage of the moderating effects of the earth, but also provide ample sunlight.

Barnard studied the idea for two years before starting construction on a 1,200-square-foot demonstration cottage. Completed in 1973, Barnard's Ecology House sits on a 50-by-50-foot lot in Marston Mills, Massa-

An interior view of John Barnard's Ecology House in Marston Mills, Massachusetts. (Photo: John E. Barnard, Jr.)

chusetts. Because the house is completely buried beneath 10 to 16 inches of earth, the only objects visible from the road are the railing which surrounds the atrium and the solar collector panels which tap the energy from the sun. Steps descend to the open-air atrium around which are a living room, kitchen, and bedroom. Floor-to-ceiling glass panels provide ample sunlight in each room, and the home stays brightly lit even on the cloudiest days.

"I would not be happy in a totally underground house," says Barbara Barnard. "But as far as I'm concerned, with an atrium like this, it's ideal. I love it."

Each summer, thousands of visitors to Cape Cod stop in Marston Mills and inspect Ecology House. Barnard's come-and-see policy is designed to familiarize the public with a radically new but practical concept for living. Barnard's purpose in constructing the home was to show the viability of the underground atrium design.

When Ecology House was first built, Barnard's neighbors and friends were convinced that he had embarked upon architectural madness. But time and ever more expensive utility bills have tempered such harsh opinions. By designing a practical home which can trim 25 percent from construction costs and at least 25 percent from fuel costs, Barnard ranks as one of those rare visionaries who appears at the right time with the right idea.

There were other pioneers, men and women who had the foresight to embrace the earth-shelter option long before it became fashionable. But

these five architects provided the lion's share of the impetus to the budding earth-shelter movement. Their underground creations were in place
when the OPEC oil embargo first emblazoned the phrase "energy crisis"
on the American consciousness in 1973. And while building underground
boasted a variety of unique advantages, it was the enormous potential for
energy savings which transformed earth-shelter architecture from a novelty into a legitimate branch of progressive architecture.

4

The Energy Crisis
and Architecture's
Changing Role

Although the human body can withstand a wide variation in temperature and humidity, the so-called zone of comfort—the ideal range of temperature and humidity—lies within a comparatively narrow band. Of course, the zone varies from country to country, person to person, and one time of the year to the next. Generally, this comfort zone for Americans ranges from sixty-eight to seventy-eight degrees F. and 20 to 80 percent relative humidity.

Theoretically, if the temperature and humidity would adhere to these limits, we wouldn't need buildings; we would erect a tarpoline to block the rain and keep a few paperweights handy to prevent the grocery list from being blown away by the wind. Yet no region of the country—not even sun-kissed Hawaii—boasts such a utopian climate. And even if it did, we'd still need buldings for security and privacy. Solid walls also provide protection from such human frailties as light fingers and prying eyes. In short, as long as man continues to inhabit the planet, there will be a pressing need for buildings and architects to design them.

According to the dictionary, architecture is the art and science of designing buildings. It is an ancient profession, one with a rich legacy of outstanding achievements: triumphs such as the Parthenon, Colosseum, Angor Wat, and Taj Mahal. But only a handful of the world's buildings are well known, earmarked by their superior design; most structures are decidedly more mundane. For this reason, we tend to take architecture for granted, surrounded as we are by a plethora of anonymous buildings.

Nevertheless, architecture has always exerted a profound but subtle influence upon human existence. The crowded tenement, the soaring sky-scraper, the sprawling ranch home nestled quietly in the suburbs—each of these buildings adds or detracts to the overall quality of life. Each helps shape behavior, moods, and attitudes. Buildings mirror a culture. They reflect a society's mores, its feelings about itself and its members, its aspirations for the future. To know a culture, one must examine its buildings as well as the people who occupy them.

What does an examination of twentieth-century architecture tell us about contemporary American society?

Ever since the 1920s, American architecture has been dominated by what is commonly called the international style. Such buildings are char-acterized by the absence of ornamentation and a sleek, uncluttered de-sign. Some of these structures are esthetic masterpieces, worthy candi-dates for architectural recognition. Yet more often than not, this style has produced sterile, unimaginative shoeboxes that are as synonymous with our culture as fast-food hamburgers, artificial turf, and throwaway dia-pers. Disdaining customs, tastes, and local tradition, the international style has resulted in buildings which are largely indistinguishable from one another. From one city to the next—and from one corner of the world to the other—high-rise boxes tower above the horizon, flaunting their clonelike lack of diversity.

The harbinger of this architectural wasteland was technology. At the beginning of the twentieth century, architecture was profoundly in-fluenced by a variety of new developments. One was the elevator. No longer were buildings confined to a height which people could negotiate by stairs. Thanks to Otis's invention, a building was free to soar as high as the architect's imagination. New building materials such as steel, glass, and concrete also appeared, destined to alter permanently the architect's craft. And new ideas were afloat. Emphasizing economy through monot-ony, the concept of mass production appealed to architects with a utilitar-ian view of their profession.

Before long, tradition architectural styles—Greek revival, Roman, Renaissance, Early American—were abandoned. Architects embraced a new god: the machine. Buildings became "functional." Frill was destined to follow the horse and buggy into oblivion. Architects turned to the clean, precise movements of the factory for inspiration. The building be-came a machine, one to which its occupants were forced to adapt. And the inevitable standardization and monotony of modern architecture was, to its proponents, as much an evidence of its success as Ford's bustling assembly lines.

Wedded to technology, the buildings of the international style turned to mechanical systems to achieve satisfactory human comfort levels. Although today we take such elaborate, energy-gulping systems for granted, their marriage with architecture was a comparatively recent event. For centuries, the only way man could heat a building was to in-

stall a fireplace or stove in each room. Cooling a structure was simply un-
heard of.

In the mid-nineteenth century, crude heating and ventilation systems were developed. Powered by steam, these mechanical marvels eventually became a standard feature of most buildings. Energy was then a
relatively inexpensive commodity, and these systems proved enormously
popular with building occupants, providing far more warmth and comfort
than the stoves and fireplaces they replaced. The invention of air conditioning in 1911 completed the cycle. By addressing the age-old problem
of cooling, it allowed architects and engineers to circumvent the weather
and provide year-round, mechanically generated comfort levels.

What evolved was the sealed building. Oblivious to its environment,
free and independent of nature's whims, the sealed building thrived on
its technological virtuosity. Too much sun streaming through west-facing
windows? Just activate the air conditioning on that side of the building.
Not enough light? Flick a switch and an entire floor becomes illuminated.
Too much heat loss through glass panels? Adjust a thermostat and it's
warm again.

The architects of the international style were free to scoff at site location, to disregard the movement of the sun, to snub their noses at environmental factors hitherto beyond their control. Teamed up with the
ever-growing army of technocrats, they became convinced that technology—and the massive quantities of cheap energy to fuel it—would liber-
ate architecture from the climatic constraints which had plagued the profession for so long.

This unquestioning reliance upon engineering was never more
amply demonstrated than in New York's World Trade Center. Dedicated
in 1973, these twin towers soar 1,350 feet above lower Manhattan. On an
average weekday, they consume as much electrical energy as a city of
77,000. Hermetically sealed from the ebb and flow of the world around
them, these 110-story energy guzzlers symbolize what may well be
remembered as one of the last triumphs of architecture over common
sense.

Long before complex mechanical ventilation systems came into
vogue—in fact, long before the fireplace and the stove became staples of
castles and cottages—man was able to design buildings uniquely adapted
to their environment. Today, such structures are called vernacular architecture. What distinguishes them is that instead of ignoring daily and sea-
sonal climatic fluctuations, they exploit them. Problems in building design are solved by paying close attention to local climatic conditions, local
building materials, and local needs. Sites are carefully selected to take
maximum advantage of the sun. What results is a remarkably practical,
amazingly efficient structure which requires little if any additional energy
to achieve adequate levels of human comfort.

Consider the following examples:

- For more than five hundred years, villagers in Hyderabad, Pakistan have employed wind scoops called *bad-gir* to cool their homes. Located on roofs and pointed in the direction of prevailing breezes, these scoops trap the cool afternoon winds to ventilate individual rooms.

- In the African nation of Chad, beehive-shaped homes built entirely of mud absorb heat during the day and reradiate it slowly at night. Although these homes are esthetically unappealing by Western standards, they are an eminently practical solution to a classic architectural problem. They consume no energy.

- The Alaskan igloo is yet another remarkable example of adaptation to local climate. Built from blocks of snow, the igloo is glazed with a sheet of ice, which forms as a result of body heat and the heat generated by oil lamps. Thanks to this glaze, the igloo is not only strong and windproof, but also well insulated. Even when outside air temperatures dip to minus thirty degrees F., the inside temperatures remain a comfortable sixty-five degrees.

Despite its unassailable practicality, vernacular architecture has been largely relegated to history texts. Nonetheless, Western architects often find it difficult to duplicate the successes of these low-tech, vernacular endeavors, and contemporary architecture can never match the rich diversity of these more primitive structures. To many of today's architects, vernacular buildings are an interesting curiosity, but old-fashioned and decidedly antimodern.

The growing preponderance of high-tech buildings paralleled an American trend toward ever-higher energy consumption. During 1955–1975, the golden age of American technology, the population of the United States increased by 33 percent. During the same period, electrical energy use leaped 400 percent.

In the beginning, there wasn't much to worry about. Until the 1950s, the United States had been energy self-sufficient. American oil fields were brimming with liquid gold, and there was a new source of energy on the horizon: nuclear fission. Early attempts at harnessing the atom proved so successful that some optimistic scientists suggested that the day was near when nuclear energy would be so cheap and plentiful that utility companies wouldn't bother to meter it.

The once-bright promise of nuclear energy never managed to live up to those heady expectations, and America's insatiable demand for energy eventually began to deplete domestic reserves of petroleum. During the 1960s, the United States was importing approximately 20 percent of its oil from foreign sources. By 1973, the figure had jumped to 36 percent. What began as a dependency had become an addiction.

Then came October 6, 1973: Egypt and Syria launched an attack on Israel. To prevent an Israeli defeat, the United States intervened and resupplied Israel with weapons and materiel. Although the Arab states lost the war, they did not take their defeat lightly. On October 19, they an-

The twin towers of the World Trade Center in New York consume as much energy as a city of 77,000 people. (Courtesy: New York Convention and Visitors Bureau.)

nounced an embargo on exports of crude-oil shipments to the United States.

The surprise curtailment of OPEC oil to the United States sent shock waves reverberating throughout American society, resulting in gas lines, closed factories, and increased unemployment. After the five-month embargo was lifted, Americans received another jolt: preembargo oil, which had sold for $2.40 a barrel, increased 470 percent in one year.

It can be argued that in the long run OPEC did Americans a favor. By disrupting our lives and tapping our pocketbooks, the oil-exporting nations put consumers on notice that the days of cheap, boundless energy were over. And just as this realization became engrained in the American consciousness, so, too, did the phrase "energy conservation" make a lightning entry into the American vernacular.

Not surprisingly, this phrase began to echo throughout the architectural community as well. Buildings consume nearly 40 percent of the total United States energy supply. Much of that energy is wasted because of unnecessary lighting, inefficient heating and cooling, and poor design. Clearly, the role of the architect was crucial if the United States were to achieve energy self-sufficiency.

But the majority of American architects were less than enthusiastic about hopping aboard the energy-conservation bandwagon. Many perceived the energy crisis as a sham engineered by the oil companies to increase already enormous profits. Others, although convinced they should conserve, discovered they were poorly trained to effect any real energy savings in design. Architecture schools, trade journals, and the American Institute of Architects (AIA) responded to this pressing need for up-to-date conservation data. But this belated shift toward an energy consciousness came only *after* the OPEC oil embargo had starkly highlighted the bankruptcy of contemporary architectural design.

Some architects found energy-saving salvation by rededicating themselves to technology. Computers were programmed to monitor heat loss; office buildings were divided into zones with elaborate closed-loop feedback mechanisms; flat-plate collectors were touted by solar-energy buffs; heat pumps and heat-reclaim systems were proposed as modern add-ons to contemporary ventilation systems. Other architects advocated a less complex approach to energy savings by relying upon better insulation and storm windows to trim heat loss during winters.

Both approaches offered valid methods of achieving substantial energy savings on existing structures. Yet as late as 1980, the majority of American architects continued to overlook the most obvious solution of all: a more energy-conscious design of new structures.

Those architects who embrace the new energy ethic tend to be a special breed. Comparatively young, environmentally attuned, and relatively unshackled to conventional architectural dogma, they preach a simple yet fiery brand of conservation ethic: architects can no longer ignore local climatic factors or the subtleties of site location. Architects can no longer build identical buildings in Phoenix and Buffalo. Architects can no longer design sealed boxes which gulp oceans of fuel. Energy is in short supply, and the architect has a duty to conserve ever-dwindling reserves. The role of the architect and architecture must change.

Had energy remained as cheap and plentiful as it was in the early 1970s, earth-sheltered architecture would still have offered a host of attractive options. But the energy crisis and the resulting shift in attitudes among more progressive architects only heightened interest in the underground alternative. The well-built subterranean design offers energy savings which few surface dwellings can duplicate.

5

Dirt-covered
Energy Savers

If earth-shelter enthusiasts tend to get carried away talking about the merits of underground buildings, forgive them. Like most converts, they are a bit fiery in defense of their new-found cause. Prior to their conversion, today's earth-shelter advocates didn't realize they had an alternative to surface dwellings. Once they discovered the practicality of going underground, they became determined to share their discovery with everyone else.

To understand why building beneath the earth can result in energy savings of 75 percent or more, consider the following comparison:

Suppose you live in a conventional brick surface dwelling somewhere in the northern Midwest. Like most people, you are most comfortable when the temperature ranges between 68 and 78 degress F. Each year, the temperature outside your home varies from approximately 95 degrees F. in summer to −20 degrees F. in winter, a spread of 115 degrees. To remain comfortable, you either have to heat or cool your home whenever the ambient air temperature falls above or below your comfort zone.

During winter, two culprits rob your home of heat. The first is *infiltration.* Your home may look tightly sealed, but it is not. Air enters and escapes through cracks around doors and windows; other cracks in ceilings, basements, and frames provide additional passageways. And because heat always flows from warmer areas to colder ones, you're liable to encounter a significant heat loss as a result of infiltration. In effect, when

Even a solid, well-built surface home such as this one can't match the energy efficiency of an earth-sheltered building. (Photo: David Martindale.)

you turn up the thermostat, you're not only warming your home, but a portion of the outside air as well. Strong winds only increase the rate of infiltration.

The second way in which your home loses heat is by transmission through the walls. Called *thermal conduction*, this slow molecular process transfers heat from one body to another. During winter you not only heat the air within your home, but also, because of thermal conduction, the brick walls. This same phenomenon occurs on the outside of your home also, bleeding heat from the brick walls to the surrounding air mass. The colder the day, the greater the heat loss.

In the summertime, heat loss is replaced by heat gain. Infiltration and thermal conduction pull heat into your home, sending you rushing for the air conditioner. Energy-saving devices such as weatherstripping, storm windows, and better insulation can help cut down on heat loss and heat gain. Even the most poorly designed home can realize substantial reductions in energy use as a result of these improvements. But because of its vulnerable location on the surface, your home must wage a constant battle with nature and its 115-degree onslaught.

Not so underground.

The reason: the temperature of the earth does not vary as much as the ambient air temperature. Dig just eight inches beneath the surface and you've practically eliminated wide fluctuations. Dig down ten feet and the temperature varies not daily, but seasonally. It is not that the earth is a good insulator—three feet of earth have the same insulation value as three-quarters of an inch of polyurethane. But the earth is an excellent temperature moderator. A 1,600-square-foot building surrounded by three feet of earth has 300,000 BTUs of heat-storage capacity wrapped around it. It takes a long time to heat and cool an 800-ton mass of earth that size. And this same mass also protects a building from expansion and contraction, as well as freeze-thaw damage.

Because of the so-called thermal-flywheel effect, the temperature of the earth lags several months behind the extremes above ground. The harsh assault of winter will cause the earth to reach its coldest point just about midspring, and it doesn't reach its maximum heat until those brisk autumn nights in October.

Even at its most severe temperatures, the earth offers a far gentler range than those found topside. At a depth of ten feet beneath Minneapolis, the yearly variation in temperature ranges from forty-seven to fifty-one degrees F. If you lived in an earth-sheltered home in Minneapolis in the winter, you would only have to heat your home twenty degrees to achieve adequate levels of comfort. Thus heating units in earth-sheltered buildings don't have to be nearly as large or as wasteful as their surface counterparts. Even without heat, an earth-sheltered building will lose only about one degree centigrade per day during winter, compared to a heat loss of about two degrees centigrade *per hour* in a conventional surface dwelling.

Clearly, the earth affords excellent energy savings, especially in climates with extremes of heat and cold. An earth cover also drastically reduces the problem of heat loss and heat gain from infiltration. Since there are far fewer surfaces exposed directly to the outside air, there is less opportunity for heat to escape in the winter and enter during summer. A properly sited earth-sheltered building also offers a formidable barrier against prevailing winds which abet infiltration. As for thermal conduction, the relatively mild temperatures behind the walls of a subterranean structure means a slower rate of heat transfer than on the surface.

Earth-sheltered buildings offer a diversity of design far richer than most conventional structures. Yet although they frequently differ in particulars, earth-sheltered structures by their very nature, share several common characteristics which set them apart architecturally. The following is a brief primer on this underground phenomenon.

SURFACE RELATIONSHIP

Earth-sheltered buildings are categorized into four basic types. As defined by architect Kenneth Labs, these include:

Chamber

This is the design pioneered by Jay Swayze, in which the entire building is buried. Only an entranceway penetrates the ground cover and provides contact with the surface. Because of this isolation, the chamber design offers an extremely energy-efficient environment. However, since it lacks natural sunlight, the chamber holds little appeal for home owners and office workers. A far more practical use of the chamber design is for buildings such as theaters, museums, laboratories, and warehouses—structures which often are built purposely without windows.

Atrium

This is the concept employed by John Barnard in his Ecology House. Completely below grade and built around a sunken courtyard, the atrium design trims energy costs and permits ample sunlight to bathe the interior rooms. The biggest drawbacks are internal circulation patterns (especially in residential buildings) and the potential for flooding in vulnerable areas.

Elevation

The Winston House is an example of this design, in which the earth-sheltered structure is cut into the slope of a hill. Three sides are buried beneath an earth cover, and one side of the building, usually the side facing south, is exposed to the sun. Additional light is gained from skylights which can be covered with insulated shutters at night to lessen heat loss.

Bermed
(also referred to as *penetrational*)

A structure built on-grade but topped by a hill, the bermed design permits windows to face in any direction. The more windows in such a building, however, the greater the opportunity for heat loss.

When viewed from inside, most earth-sheltered buildings look remarkably similar to their surface counterparts. Rooms are rectangular, roofs either flat or sloped. Although shelltype designs like the Clark-Nelson home offer great promise because they can support heavy earth loads, they have yet to be adapted on a wide scale, primarily as a result of their unusual shape.

CONSTRUCTION MATERIALS

Just as most earth-sheltered buildings look like conventional structures from within, they are also built with the same conventional materials.

Concrete

By far the most popular and versatile building material for earth-sheltered structures, concrete is used for floors, roofs, walls, columns, and beams. Although durable and fire resistant, this material is not without its drawbacks. Untreated concrete tends to shrink and crack when dry, causing leaks. Concrete blocks are relatively porous and too weak to withstand great earth pressure. For this reason, many earth-sheltered structures are built with reinforced concrete or concrete which has been posttensioned, a process which makes the material stronger and more watertight. In addition, precast concrete planks are often provided with steel reinforcement, which makes them ideally suited to supporting heavy earth loads.

Steel

Because of its great strength and ability to remain watertight, steel seems a logical first choice for most underground construction. But steel has one major drawback: it is expensive. For this reason, its use for beams and columns is usually confined to large commercial structures rather than homes.

Wood

Comparatively strong, easy to work with, and relatively inexpensive, wood is often used in the familiar post-and-beam systems as well as on interior walls. Its rich earth tones are ideally suited for a design whose very origin is the earth. But the use of wood also poses a problem. Theoretically, an earth-sheltered building is designed to last forever; wood is not. Chemical preservatives can retard decomposition, but no one is certain if such chemicals prevent or merely postpone decay.

No matter what the construction material, it takes engineering know-how to properly design an earth-sheltered building. Even though these structures are relative newcomers to the architectural scene, the often-feared structural failures or "cave-ins" of underground buildings simply haven't occurred. Ironically, roof collapse remains the sole province of surface buidings, usually those weighted down by excessive loads of rain or snow.

SOIL CONSIDERATIONS

Although earth-sheltered buildings are extraordinarily energy efficient, it is not always easy determining just how much heating will be needed for any particular structure. With a surface building, an engineer or an architect has to consult standards published by the American Society of Heating, Refrigeration and Air-Conditioning Engineers (ASHRAE). Such standards don't exist for underground buildings, and different soils have different degrees of thermal conductivity. Soil analyses of individual sites are required to determine the precise amount of heat loss which will occur.

Such analyses are also crucial in determining the location of the local water table. Although earth-sheltered buildings can be constructed on practically any terrain, a site on swampy land, a flood plain, or an area where the water table is high generally requires expensive drainage equipment and more complicated building techniques.

Despite the wide range of soil conditions, there is seldom a problem with earth-sheltered buildings settling. These structures often weigh less than the massive quantities of earth they replace.

DEPTH OF EARTH COVER AND INSULATION

The type of soil and its moisture content also determines how much earth should cover an earth-sheltered building. Invariably, the roof of an underground building provides the greatest potential for heat loss. For this reason, the deeper the earth cover, the greater the energy savings. Not only will more soil trim heat loss, but it will also permit adequate drainage and provide room for plant roots to take hold.

But digging deeper poses a major problem: the more earth which covers a roof, the greater the weight. In fact, just two feet of earth exert a pressure of 200 to 240 pounds per square foot. Not surprisingly, supporting such heavy weight quickly translates into extra costs. Therefore nearly all earth-sheltered roofs are thickly insulated. This trims heat loss and acts as a cost-efficient compromise to burrowing still deeper. Insulation is usually installed outside the upper half of the building's walls, especially on concrete structures.

WATERPROOFING

Because they are topped by a generous supply of earth, underground buildings must be adequately waterproofed. Despite the all-too-familiar problem of leaky basements, earth-sheltered structures boast an enviable record for moisture control. The reason: only the best, longest-lasting sealants are used. These range from bituthene, a rubberized asphalt with a polyethylene coating, to cardboard panels stuffed with a claylike sub-stance which expands on contact with water. As long as such products are properly installed, they are likely to last the life of the building.

HUMIDITY CONTROL

Usually, the only significant underground moisture problems occur in concrete structures on hot, muggy summer days. As moisture evapo-rates in the concrete, condensation forms on the cool walls. Generally, this condition persists for about two years. Since few underground build-ings require air conditioning, this excess humidity is controlled by a de-humidifier.

By far, the biggest heat loss in an earth-sheltered structure occurs through windows. But most architects are more than willing to tolerate this heat loss. Windows provide not only sunlight and a view of the out-side world, but also solar energy. Combined with the large thermal mass around the walls, solar energy can help a building achieve even greater energy savings.

6

The Solar Underground:
Let the Sun Shine Down

Compared with most stars in the universe, the sun which dominates our solar system is not an astronomical standout. It is smaller and less powerful than most stars. In short, it is a celestial mediocrity, a star as nondescript as a third-string pitcher in the minor leagues.

Despite its bland credentials, the sun remains a powerful energy factory. Each second, nuclear fusion converts 4.7 million tons of mass into radiant energy, a process which is expected to continue for the next 5 billion years. Only a small fraction of this radiation reaches the earth, but the sunlight that does bathe our planet is essential for human existence. The sun is the source of both life and energy.

Fossil fuels such as coal, natural gas, and oil can be traced to sunlight which illuminated the planet millions of years ago. The winds which propel our sailboats, the rains which generate hydroelectric power, and the forests which give us kindling wood have their legacy in the sun's life-giving radiation. In the broadest sense, all energy is solar energy; the sun is man's ultimate fuel.

It has only been since the advent of the "energy crisis" that solar energy has become a common phrase in American vernacular. When used in its contemporary context, solar energy refers to attempts to harness directly the power of the sun. Sometimes these efforts involve electrical generation using silicon cells, a process used extensively in space flights. But more commonly, solar energy implies the use of the sun to heat and occasionally cool buildings and homes.

The growing army of solar advocates is quick to tout the impressive advantages of the sun. Unlike fossil fuels, solar energy is inexhaustible. Solar energy is also environmentally harmless, neither polluting nor destroying. And solar radiation is immune to embargoes. "No cartel controls the sun," President Carter once aptly observed.

Solar energy also has enormous potential to help the United States solve its energy woes. Seemingly overnight, the solar option has gained tremendous popularity. In 1975, there were only 183 solar-heated homes in the United States. Before the decade ended, more than 40,000 buildings boasted some type of solar device. Despite scarce capital and frequent bankruptcies, what was once nothing more than a fledgling cottage enterprise is well on its way to becoming a multibillion-dollar industry.

Although solar apparatus vary greatly, most active solar heating systems consist of four basic parts: collectors, heat-storage system, delivery unit, and back-up system. Mounted on the roof, flat-plate collectors consist of blackened metal panels surrounded by glass and heavily insulated on the side away from the sun. The heat from the sun is trapped in the metal and then transferred either to air or water. Electrical power is used to pump the air or water through ducts or pipes to the storage unit. Heat is then delivered to the building via a conventional forced-air heating system or else as steam heat through baseboards or radiators.

Properly designed, a water-tank or rock-pile storage system can supply enough heat for two or three cloudy days; a full-sized, conventional back-up heating system is needed to provide additional warmth on several cold, sunless days. With adaptations, solar collectors can be designed to provide cooling in the summer.

A wide-scale use of solar energy to heat and cool America's buildings would go a long way toward reducing the United States' dependency on foreign oil. Yet the sunny promise of solar energy is clouded by several drawbacks. Like all high-tech ventures, the solar collector uses complex machinery. Breakdowns are not uncommon, and maintenance is a frequent headache. Moreover, solar collector panels are not energy independent; energy input is needed to produce the metal and glass used to trap the sun's heat, and additional energy is required for the collectors to function.

The biggest cloud on the solar horizon is cost. Despite their overall savings in the long run, active solar heating systems require a significantly greater capital investment than standard gas- or oil-fired heating systems. This is not to imply that active solar heating systems are doomed to the status of an expensive ecological fad; continued research and development may well lower the price of collectors in the same way the "miracle chip" has trimmed the cost of pocket calculators. If collector-panel prices drop and oil and gas prices continue to soar, solar energy will certainly become a more attractive method of conserving energy in the future.

Although solar panels remain an expensive option for most conventional buildings, they are often surprisingly affordable when used with earth-sheltered designs. The reason: most solar collectors on surface buildings cover an area roughly half the size of the floor space; a 2,400-square-foot home requires about 1,200 square feet of collectors. Yet this same rule of thumb doesn't apply underground. Because of their very location, earth-sheltered buildings require far less energy input to heat and virtually none to cool. They need far fewer solar collector panels than ordinary structures, and thus make possible a substantial savings in capital investment. Throughout the country, a growing number of earth-sheltered buildings are pointing collectors skyward and tapping the energy of the sun.

Ever since solar energy captured the imagination of the American conservationist, active collector systems have garnered the lion's share of attention, research, and publicity—so much so, that when most people hear the word "solar," they think of flat-plate collectors gleaming brightly on rooftops. However, there remains another form of solar energy available for harnessing, a form less sophisticated than its active counterpart, but one which is ideally suited to underground adaptation.

This type is called passive solar, and its beauty lies in its utter simplicity. Solar radiation penetrates a building, is stored within the building mass, and then released slowly once the sun has disappeared. This concept predates the recent vogue in solar energy: ancient Greeks and Romans incorporated passive solar design into their buildings because of a shortage of kindling wood. Homes were built facing south and designed to provide maximum solar gain in winter and minimum solar penetration during summer. Nearly all vernacular architecture utilized the sun favorably so that the building itself became a giant heat collector during cold weather. Such common-sense architecture exists in many Third World nations today.

Intoxicated by technology, modern Western architects have until recently shunned the passive solar design as outdated and unnecessary. Earlier in the century, the passive solar promise enjoyed a brief revival thanks to two architect brothers, George and William Keck. While designing the Crystal House for the 1933 Chicago World's Fair, the Kecks noted that although the temperature outside was cold, workmen constructing the glass building were laboring in shirt sleeves. Intrigued, the Keck brothers began investigating passive solar design, and over the next decade, they built several homes which maximized heat gain from the sun during cold weather.

As George and William Keck discovered, the secret of successfully utilizing passive solar energy has nothing to do with machinery. It requires intelligent building design. Proper siting must be one of the primary considerations. Exposing a large number of windows to the east or west only increases the heat gain in summer and trims it in winter. In

Built for the 1934 Chicago World's Fair, the Crystal House led architects George and William Keck to design several passive solar homes. (Courtesy: George Fred Keck–William Keck Architects.)

order to take full advantage of passive solar radiation, most windows must face within thirty degrees of due south.

Such a design also requires that the architect determine the position of the sun at various times of the year. The sun cuts across a much lower arc in winter than in summer. Depending upon the exact latitude, this difference can be as much as forty-seven degrees. Unless adequate precautions are built into the design, the same solar energy which will help heat a building during winter will also raise temperatures during summer. For this reason, architects often utilize overhangs, roof extensions positioned to block the entrance of direct sunlight in summer. A similar function is sometimes served by creative landscaping, particularly the use of deciduous trees. During the summer, direct sunlight is blocked from entering the windows by the trees' leaves. During winter, these now barren trees allow sunlight to penetrate easily into the building interior.

Most passive solar buildings make generous use of double- or triple-pane glass on the south-facing windows. This glazing not only allows solar energy to enter the building envelope, but it also traps the longer-wave heat radiation from escaping back into the atmosphere. Even during the middle of the winter, bright sunlight streaming through glass can quickly provide too much heat unless that heat is stored. This is why all successful passive solar designs incorporate a large thermal mass into the structure.

Enter the earth-sheltered building. Because of its extremely large thermal mass—in most cases, tons of poured concrete in floors, walls, and ceilings—the well-designed passive-solar underground building operates as a classic "heat sink." All day long, this immense mass slowly absorbs excess heat like a sponge soaking up water. The entire building becomes a giant heat-storage system, protected from heat loss by the insulation surrounding the walls and roof. At night and on cloudy days, the thermal mass releases this stored energy very slowly, warming the building even though the sun has disappeared.

The passive solar concept is not restricted to earth-sheltered structures. Properly designed, surface buildings can also utilize this abundant and remarkably uncomplicated source of energy. However, it is nearly impossible to adapt most existing conventional structures to the passive design. Not only are they improperly oriented, but they generally contain far too many lightweight low-mass materials which are inadequate heat collectors.

Earth-sheltered structures, on the other hand, are passive solar naturals. A huge mass is a function of the building design. For a good many underground buildings, it is downright foolish *not* to take advantage of the passive solar potential.

Baldtop Dugout, the home of architect Don Metz, clearly demonstrates this ideal marriage between earth-sheltered and passive solar designs.

Built in 1977, Baldtop Dugout, like its predecessor the Winston House, is cut into the side of a hill in western New Hampshire. Nearly all of the south wall is glass, and the interior portions of the six-room home are lighted via skylights. The seven-foot-four-inch ceiling is low enough to reduce the amount of space to be heated. Outside the home, both fin and wing walls protrude from the front of the building. They act to limit the strong winds which buffet the hilltop and to reflect the sunlight off the snow cover in the winter.

Sunlight which enters the 1,800-square-foot home is immediately absorbed by a number of high-mass materials: concrete walls, brick archways, slate counters, fir and hemlock beams, and tiled floors. Even when night temperatures dip to freezing, this high thermal mass provides ample heat to maintain a seventy-degree F. indoor temperature. On colder days, Metz uses a potbellied stove to remove the chill, burning roughly five to eight cords of wood during a typical winter. Only rarely does he use a small back-up oil furnace.

Besides being a practical, energy-efficient home, Baldtop Dugout is also an esthetically pleasing piece of architecture. Rooms are curved and

Inside Baldtop Dugout, high-mass materials such as concrete walls, slate counters, and tiled floors help store passive solar energy. (Photo: Ross Chapple.)

dominated by rich wood ceiling beams. Earth colors predominate throughout the home, and there is a noticeable variety of forms and textures. Despite its bermed walls and eight-inch earth cover, Baldtop Dugout belies its underground presence not only by encouraging the sun to brighten the interior, but also by putting that sunlight to work.

Like any form of energy, passive solar has its drawbacks. By their very nature, the necessary designs impose severe architectural restraints. Gone is the freedom to ignore nature. Buildings must be properly oriented and constructed with only top-quality, high-mass materials. Also, the passive solar concept works best in smaller buldings where stored radiant energy travels relatively short distances.

Nevertheless, for a growing number of people, the drawbacks to passive solar energy are outweighed by its advantages. Buildings such as Baldtop Dugout are the quintessence of low-tech architecture. And because they use free solar energy without elaborate mechanical devices, they are less expensive to build than structures employing solar collectors. Passive solar buildings are also silent and maintenance-free.

Perhaps best of all, a structure like Baldtop Dugout celebrates an architectural philosophy aimed at a return to basics. Long before human beings began roaming the planet, there was earth and there was sunlight. Unlike the dynamo and the oil rig, both are a natural part of the environment. Both are waiting to be tapped. It is only fitting that teamed together in a well-designed structure, they provide such a simple and effective way to conserve energy.

7

The Lower Ground
of Higher Learning

In September 1962, the Abo Elementary School in Artesia, New Mexico opened its doors. By any measure, Abo was not a run-of-the-mill elementary school. Designed to double as a fallout shelter, the 144-by-200-foot windowless building was built completely underground. On the roof, a basketball court rested atop a 21-inch thick slab of poured concrete. Inside, the building contained such special features as particulate filters, a separate internal water supply, emergency power generators, and decontamination showers.

Because of its subterranean location, educators were curious about its effect on students. In a test conducted during the first school year, researcher Frank Lutz found no evidence that Abo's windowless environment had any effect whatsoever on student achievement, behavior, anxiety, or health. Nine years later, Lutz conducted a more exhaustive follow-up study on Abo. The results were almost identical, except that there was evidence pointing to possible favorable effects of windowless classrooms on physical and mental health.

A similar study was conducted in 1962–1963 by the Architectural Research Laboratory at the University of Michigan. To determine the effects of windowless rooms upon the learning process, researchers studied students at two Wayne, Michigan elementary schools. Both schools were identical surface structures, except the test school's windows were removed and replaced by concrete. According to the final report, "the test school children have shown very little personal interest in whether their classrooms had windows or not." However, interestingly enough, teachers preferred the classrooms without windows. Not only did the absence of windows provide extra wall space which the teachers put to good use, but it also prevented students from becoming distracted by outside events.

Although Abo Elementary School didn't exactly precipitate a boom in underground schools, it was the first of several such subsurface learning environments, most of which are located in the Midwest and Southwest. Sometimes, the reason for going under was directly related to civil defense, the rationale being that if you were going to build a fallout shelter to provide protection against nuclear war, you might as well put the structure to good use during peacetime. In other cases, noise reduction was the primary purpose for choosing the underground option. Schools at Roswell, New Mexico and Lake Worth, Texas were buried to deafen the roar of military jets at nearby air bases. Still other schools exploited the earth-sheltered design as a defense against the elements; by 1980, more than forty schools in tornado-plagued Oklahoma were at least partially bermed in order to provide protection from the annual barrage of twisters.

As school officials were quick to discover, such locations provided a number of advantages, the lack of student distraction being only one. Compared with conventional buildings, maintenance costs were reduced sharply. So was the potential for vandalism. And the savings on heating and cooling were dramatic, although reduced energy consumption was a by-product of the design, not the original purpose.

Inevitably, skyrocketing energy costs had an impact on school administrators, as they did on other segments of American society. Energy costs were paramount in the decision of the Fairfax County school board to build a new elementary school in Reston, Virginia. The board's instructions to architect David Carter were specific: design the most energy-efficient building possible. Taking his charge seriously, Carter did just that. Calling his plan Terraset, he designed an earth-sheltered school which rested on ground level but was covered by soil ranging in depth from eighteen inches to nine feet.

Even though Terraset would provide ample sunlight and windows, the community's initial reaction to the plan was decidedly negative. The cave associations persisted. "Who wants to send their kids to a school where the students need mining hats?" was a typical response. Not only were parents unhappy with Carter's plan, but so was the school board and a state review agency. No one argued with the fact that Terraset would save considerable energy; the school's earth cover gave rise to the concern. Fortunately, cool heads prevailed and the $2.9-million structure was translated from the drawing board to the suburban Washington, D.C. countryside.

Opened in February 1977, Terraset provides a unique learning environment for 990 students from the first through sixth grades. A glass-

The entranceway to the earth-sheltered Terraset Elementary School in Reston, Virginia is spanned by solar collector panels. (Photo: David Martindale.)

enclosed courtyard in the front of the school serves as a drop-off point for students. Flanking the courtyard are administrative offices, a gym, and a cafeteria. Inside, the 66,000-square-foot structure is divided into four circular learning centers. Each boasts a curved window which admits generous amounts of sunlight. Partitions within the learning centers divide the space in each 100-foot-diameter circle. In the center of Terraset, a combination media room and library sits directly beneath a large skylight. The building's ceiling height was lowered from ten to eight feet in order to reduce the amount of air space to be heated and cooled.

Because of its extensive thermal mass, Terraset requires very little heating. The heat generated by lights and passive solar gain through the windows is usually sufficient to warm the building on even the coldest days. In fact, cool outside air is often pumped into the school during winter to keep temperature levels from rising too high. For most of the year, the problem is how to cool Terraset.

To reduce the energy required to power a conventional cooling unit, Terraset employs a heat-reclamation system which recycles waste heat generated by people, machines, and lights. Such waste heat is ordinarily extracted in the cooling process, but at Terraset it is recaptured, stored as hot water in large tanks, and then used for additional cooling by driving absorption chillers.

Spanning the courtyard high above the school's roof, a solar-panel collector system provides Terraset with an additional energy input. Funded by a $650,000 grant from the Al Dir Iyyah Institute of Saudi Arabia, the panels cover 4,822 effective square feet. Water in glass tubes passes through heat exchangers, which transfers the solar heat to the school's hot-water supply. This heat is either used directly or converted by refrigeration units into cooling. In all, active solar energy accounts for about one-third of Terraset's total energy savings.

Terraset's entire heating and cooling load is monitored by a computer programmed to trim utility costs as much as possible. Despite the fact that both the heat-reclamation system and the active solar collectors require electrical input in order to function, the savings have been substantial. Because of its earth-sheltered design, Terraset uses fully 60 percent less energy than comparably sized surface schools in Fairfax County.

Terraset is also an esthetically pleasing addition to the surrounding landscape. Blending in with the trees and foliage nearby, Terraset is a classic example of what Malcolm Wells calls "gentle architecture." In warmer weather, teachers conduct outdoor classes on the grass above the school, and the roof doubles as a playground year round.

For the students, Terraset is more than just another school. It is a day-to-day laboratory of energy conservation, a testament to responsive architecture and creature design. The early concerns about the school being buried in a hole have long since vanished; Terraset is now the jewel of the Fairfax County public school system. The earth-sheltered design is

While some children study below, others play near a skylight on Terraset's grass-covered roof. (Photo: David Martindale.)

so popular and successful that the school board plans at least one additional Terraset-style school, largely as a result of public insistence.

Elementary schools like Terraset aren't the only educational institutions embracing the earth-sheltered option. American universities and colleges have been doing so for more than a decade. One of the earliest pioneers of subgrade design is the University of Illinois at Champaign-Urbana.

In the mid-1960s, the University of Illinois library was severely overcrowded; the school needed an additional library facility. But the university faced a problem: where to put it? Constructing a conventional structure next to the original building was out of the question; it would not only have marred the appearance of the campus mall, but it would have also shaded Morrow Plots, the oldest experimental soil project in the United States. Equally unsatisfactory was a library addition separated from its parent building by a great distance.

To solve their dilemma, campus officials chose an earth-sheltered design submitted in a national competition. Completed in November 1969, the University of Illinois undergraduate library is a two-story struc-

ture surrounding a central atrium. A landscaped plaza rests atop the roof. The building is joined to the main library by a tunnel, and future underground expansion could double the size of the existing facility.

For the University of Illinois, going underground was a site decision. Similar site decisions prompted construction of underground facilities at other crowded campuses. In the late 1960s, administrators at Northern Iowa University wanted to build a student union in the center of the campus. Since the only space available was a park which officials wanted to preserve, they built the student union beneath the park. Faced with similar campus building congestion, Cornell University opted for an earth-sheltered campus store, and the University of Houston buried its student center in the heart of the campus.

More recently, two other building-choked universities have designed earth-sheltered facilities. Each is unique for a different reason.

Completed in late 1975, Harvard's Pusey Library is a three-story structure, 85 percent of which lies below Harvard Yard. Light for the 87,000-square-foot building enters from two sources: a "light moat" which surrounds the outer edges of the top floor, and an atrium which penetrates all three levels. Designed to house rare books, maps, and manuscripts, the building's foundation rests a full eleven feet below the water table. To prevent any intrusion of moisture, engineers wrapped the library in ironite waterproofing and then rigged up a complex system of drainage pipes, gravel beds, and four sump pumps. A similar system would be used should facilities at Pusey Library be expanded underground.

In Washington, D.C., the fashionable Georgetown neighborhood has long been synonymous with elegant homes and expensive real estate. For this reason, Georgetown University officials were hard pressed to meet the cost of purchasing additional land for a badly needed recreation center, nor was any land available on the already tightly congested campus. In 1977 the decision was made to tear up the existing football field, excavate a new recreation center below ground, and then reconstruct a synthetic turf field on the roof. A patented engineering technique was employed to enclose the 142,000-square-foot structure which now houses handball courts, squash courts, twelve multipurpose courts, batting and golf driving ranges, lockers, and a pool. Although construction costs were $500,000 higher than for a conventional building, the university was spared the expense of purchasing additional real estate. Best of all, Georgetown's new recreation center will trim energy consumption by two-thirds compared to a comparable surface structure.

Not all earth-sheltered campus buildings are necessarily more ex-

A partial view of the lobby of the University of Houston's earth-sheltered student center. (Courtesy: Goleman and Rolfe Architects.)

Only the entranceway to Georgetown University's Yates Field House is visible. The rooftop railing surrounds a football and track field. (Photo: David Martindale.)

pensive to construct. Williamson Hall cost an estimated 3 to 5 percent less than a similar surface structure. But energy, not construction cost, is why this University of Minnesota bookstore and office building has received so much attention. No other underground structure in the United States has received a more thorough "energy physical" than Williamson Hall.

Energy was not of paramount importance when the building was designed in 1972. Although architect David Bennett was asked to comply with the university's energy-conservation guidelines, the primary reason for going underground was a site decision. University officials were eager to preserve what little open space remained on the crowded Minneapolis campus. They wanted an unimpeded flow of pedestrian traffic which bisected the site. They also instructed Bennett to preserve the view of two historic campus buildings which fronted the site on the north and west.

What evolved was a two-story, 83,000-square-foot structure, 95 percent of which is below grade. Bennett chose to separate the bookstore and admissions and records office by slicing the 250-foot square site into two

equal triangles. A grade-level pedestrian walkway and an underground walkway cut between the two areas. Mezzanine-level bookstore offices overlook the main sales area, which lies on the bottom floor of one triangle. An above-grade concrete slab provides clerestory windows which peer into the sales floor below.

A different approach was used with the admissions and records triangle, which Bennett designed in an L-shape around an atrium. Glass slanted at a forty-five-degree angle permits sunlight to penetrate the admissions and records area during winter. Strategically placed ivy planters prevent solar heat gain during summer. Inside, brightly colored divider panels and natural birch provide a pleasing balance to rough-hewn concrete columns and walls.

Because it is nestled underground and because of the large number of people who use the building, Williamson Hall requires much more cooling than heating, despite the fact that Minneapolis is notorious for its bone-chilling winters. Williamson Hall retains so much heat from lights, machinery, and people that it requires no heating until the temperature dips below minus four degrees F. Even when the heat is shut off completely during winter weekends, the temperature inside never dips below fifty degrees F. Instead of a conventional fan-and-duct system, the building employs air fountains to distribute fresh air. In addition, heat exchangers are used to recycle waste heat, which is then used to assist in cooling.

Two years after the reinforced concrete building was opened in 1977, Williamson Hall added a panel of concentrator solar collectors to further assist trimming energy needs. Had the building been constructed above grade, fully four acres of collectors would have been needed. Thanks to its earth-sheltered design, collector panels cover an area totaling only 8 percent of the total square footage of the bulding. They provide enough energy to supply roughly half the heating and cooling load of the building.

If energy consumption at Williamson Hall receives a high priority, it is largely the result of the University of Minnesota's Departments of Civil and Mineral Engineering and Mechanical Engineering. Both departments were early and avid proponents of the earth-sheltered design. Both were convinced that the construction of Williamson Hall provided a useful opportunity to monitor the energy efficiency of an earth-sheltered building. The university received a $240,000 grant from the National Science Foundation to conduct a variety of exhaustive tests. The purpose of these studies: to make a precise determination of the energy loss through walls and floors.

Computers take periodic temperature readings inside Williamson Hall and within the walls and floor. This information is then combined with the findings of temperature probes buried in four types of soil surrounding the structure. Such data provide university engineers with a continuous diary of the building's energy performance.

As engineers are quick to point out, Williamson Hall is not an ideal earth-sheltered structure from an energy viewpoint. Some portions of the building are more difficult to heat and cool than others. Engineers badly underestimated the heat loss which occurs through the grade-level concrete slab above the bookstore. Were this slab replaced by a ground cover topped with grass or shrubbery, it would not only provide better insulation, but it would also look far more attractive.

Nonetheless, Williamson Hall works. It saves energy and it preserves open space, historic views, and pedestrian access. Although it might appear just another campus building, it is a working energy laboratory. What University of Minnesota engineers learn from Williamson Hall will ultimately be reflected in other earth-sheltered structures, both off campus and on.

8
A Potpourri of Subterranean Selections

The Soviet Union's pavilion at Expo '70 in Osaka, Japan was easy to spot from a distance. The tallest exhibition hall on the fairgrounds, the building was painted a bright red and featured a sharply curved roof shaped like a slalom run. At the very top of the structure was a prominent hammer and sickle.

The United States' pavilion, on the other hand, was a bit more understated; it was partially buried underground. The size of two football fields, the three-story earth-bermed structure enclosed a total of 226,800 square feet. The translucent roof was made of a fiber-glass membrane, supported by air pumped by four large compressors. Designed by Davis, Brody Associates, the building was dubbed "the most daring structure at Osaka" by *Architectural Forum.*

The American pavilion was one of the first nonresidential, noneducational applications of the earth-sheltered design. Though architecture critics generally lauded the American pavilion and panned the Russian one, the marriage of earth-bermeds and air-inflated structures has not caught on in the years since Expo '70.

Today, a variety of commercial and public buildings are benefiting from the earth-sheltered option. Some are comparatively small structures; others, massive. Because each is integrated with the earth around it, these buildings are able to realize impressive energy savings.

The U.S. Pavilion at Expo '70 in Osaka, Japan was buried partially underground. (Courtesy: Davis, Brody and Associates Architects.)

FLORIDA STATE MUSEUM

Completed in 1970, the Florida State Museum is located on the campus of the University of Florida at Gainesville. The building houses a natural history collection and combines display areas and research facilities. Carved into a hillside, the three-story structure was one of the first permanent public buildings in the United States to tap the earth-sheltered resource.

The Florida State Museum is an L-shaped, terraced structure com-

The earth-bermed Florida State Museum in Gainesville was designed by William Morgan. (Courtesy: William Morgan Architects.)

prising 102,000 square feet. The exhibit area and visitor entry are on the top level. Staff offices occupy the floors below. An upper-level bridge leads to an earth-bermed pyramid which descends to a sculpture garden. Designed by architect William Morgan, the building's bold, dramatic appearance is accentuated by its earth-bermed design and its prominent concrete canopies.

The museum is also an energy saver. Built a few years before the energy crisis, the building is temperature controlled to maintain a near-

Surrounded by earth and topped with clerestory windows, St. Benedict's Abbey in Benet Lake, Wisconsin barely resembles a conventional church. (Photo: David Martindale.)

constant seventy-one degrees F. When fuel prices soared, the staff was able to reduce energy consumption by 48 percent and still stay within one degree F. of the original design temperature. In large part, this was because of the building's earth-sheltered design.

ST. BENEDICT'S ABBEY

It hardly looks like a church, but then it is not supposed to. Located in the small community of Benet Lake, Wisconsin, St. Benedict's Abbey sits beside a large Tudor-style monastery. Rather than have the old building compete for attention with a striking contemporary design, architect Stanley Tigerman chose the earth-sheltered alternative. The result is a unique, unpretentious abbey, one which provides exactly the kind of low-key profile that the monks of St. Benedict's wanted.

At first glance, the church appears to be an extension of the landscape, the lawn ascending the earth-bermed walls halfway to the top of

Mutual of Omaha's glass-domed underground annex is buried in front of the firm's Home Office building. (Courtesy: Mutual of Omaha Photo.)

the building. The metal roof features a ring of triangular clerestory windows which provide a diffused light source for the interior below.

Completed in July 1972, this unassuming sixty-eight-foot-long church holds 300 worshippers, all of whom sit on simple office-style chairs beneath overhead beechwood trusses. A plain wooden pulpit rests atop a small raised platform. An enclosed bridge from the monastery leads to a balcony at the rear of the abbey from which parishioners descend onto the main church floor.

MUTUAL OF OMAHA ANNEX

Perhaps as well known for its sponsorship of Marlin Perkins's television show "Wild Kingdom" as for its role as a major insurance firm, the Mutual of Omaha corporation employs 5,100 workers at its Nebraska headquarters. By the mid-1970s, office space was becoming cramped; the firm's three existing buildings were inadequate to house its expanding staff. For this reason, executives of the company decided to build yet another annex, this one underground.

There were three basic reasons for Mutual of Omaha's decision to build an earth-sheltered structure. To begin with, the choice of nearby available land was limited. The only site which made sense was a parking lot opposite the north side of the twelve-story Home Office building. Company officials did not want to obstruct the view of this building, nor relinquish the valuable parking area.

Secondly, Mutual of Omaha planners were concerned about saving energy. The fact that an underground building could generate ample heat from lights and people meant that the new structure wouldn't need a mechanical heating system. Instead, it could tap the company's central system for what little additional heating or cooling would be required.

Finally, going underground was considerably cheaper than building on the surface. Mutual of Omaha estimates the firm saved $5 million by building below grade. Additional savings are expected in the future through reduced energy use.

Completed in the spring of 1980, the 184,000-square-foot building consists of three levels. The bottom two floors provide much-needed office space. Although these lower levels lack windows, company officials insist this causes little employee dissatisfaction. The reason: because of the size of the Home Office building, many employees in that high-rise structure aren't able to see outside either.

The top floor of the annex features a 1,000-person cafeteria, kitchen facilities, library, visitors center, lounge, and personnel-training center.

rees grow inside the garden court of Mutual of Omaha's glass-domed *rth-sheltered annex.* (Courtesy: Mutual of Omaha Photo.)

The Arden Hills and Shoreview office of the Roseville State Bank is an excellent example of a small earth-bermed building. (Photo: David Martindale.)

An indoor garden court rests beneath a huge geodesic dome which is dramatically lighted at night. This dome is all that is visible from the surface of the massive structure. Measuring ninety feet wide and fifteen feet high, the dome is made of twin layers of tempered plate glass designed to reflect the sun. The dome weighs a hefty thirty tons, and architects Leo A. Daly Company estimate it is capable of withstanding a snow load as heavy as thirty pounds per square foot.

Although the bottom floor of this earth-sheltered office building rests forty-five feet deep and below the water table, superior waterproofing and a sump pump system protect the structure from moisture. As a result of the building's superior energy performance, Mutual of Omaha expects to save 80 percent on its heating and cooling expense for the domed annex.

ARDEN HILLS BANK

Not all commercial earth-sheltered buildings occupy a great deal of space. In suburban Minneapolis, the small Arden Hills and Shoreview office of the Roseville State Bank sits unobtrusively alongside a major intersection, blending with the surrounding landscape. Bermed walls rise up to cover the precast concrete structure, which rests atop ground level. Should more space be required in the future, the building is designed to be easily expanded either vertically or horizontally.

PLANT SCIENCE BUILDING
OF THE CAREY ARBORETUM

This $2.6-million reseasrch lab in Millbrook, New York was designed by architect Malcolm Wells and engineer Fred Dubin. The Plant Science Building of the Carey Arboretum employs not only a number of common-sense conservation techniques, but also the most sophisticated computer technology to trim energy needs to a minimum.

Three-quarters of the two-story building has either been earth bermed or buried below ground. Small hills shield the timber-roofed structure from the wind. Heavily insulated, the building's large southern exposure is ideally suited to take advantage of passive solar gain in the winter. During the summer, overhangs, sunfins, and movable insulating panels block unwanted solar radiation.

In addition to its passive solar features, the 28,000-square-foot building also boasts sevens rows of active solar-collector panels, featuring 5,650 square feet of flat-plate collectors. A combination water and antifreeze liquid circulates through pipes absorbing heat, which is then stored in two 7,500-gallon tanks. To further lessen the heating load, warm air exhaust from machinery is captured, pumped to heat exchangers, and used to reheat incoming air.

This entire network is connected to a computer programmed to se-

Seven rows of solar collectors cover the roof of the Carey Arboretum's earth-sheltered Plant Science Building in Millbrook, New York. (Courtesy: Carey Arboretum.)

lect one of twelve modes for maximum energy efficiency. As a result, the Plant Science Building's heating requirement is just 20 to 33 percent that of a similar structure built on grade.

MINNESOTA'S REST STOPS AND RANGER STATIONS

When it comes to using underground space, Minnesota is one of the most progressive states in the country. Consider the following two examples:

Minnesota's Department of Transportation has constructed three earth-sheltered rest stops for motorists. Two wood-and-limestone structures are located near Blue Earth off Interstate 90; a third was built along

a major highway near Anchor Lake. All three feature observation decks as well as toilet facilities and space for machines.

The Minnesota Department of Natural Resources and the Minnesota Housing Finance Agency have jointly funded a prototype program resulting in three earth-sheltered park manager houses. Located at Whitewater, Camden, and St. Croix Wild River State Parks, these structures range in size from 1,450 to 1,700 square feet and cost approximately $100,000 apiece. The three ranger stations were designed to take advantage of passive solar gain in the winter and the Whitewater and Camden houses also feature active solar-collector panels. The University of Minnesota's Underground Space Center will monitor the thermal sensing equipment which has been built into all three structures.

FORT WORTH CENTRAL LIBRARY

In the spring of 1972, voters in Fort Worth, Texas approved $6.8 million in bonds for the construction of a central library on a two-block downtown site. Since the location was bisected by a major thoroughfare

An earth-sheltered rest stop near Anchor Lake, Minnesota. (Courtesy: Damberg & Peck Architects.)

(which city officials were reluctant to close), original plans called for the building to span the roadway, a bridge linking the portions of the structure.

Before the project got underway, a lawsuit contested the legality of the bond election. For three years, the courts heard arguments on the case. By the time architects Parker Croston Associates were given the go-ahead on the project, inflation had eaten away 25 percent of the construction budget. No additional money was appropriated for the building, and the City Council refused to trim the scale of the project. Unable to build the library above grade and stay within the budget, the architects decided to dig underground. The savings to the city as a result: more than $2.7 million.

The Fort Worth Central Library is an 11,600-square-foot poured-in-place concrete building, 80 percent of which is below grade. A surface plaza level contains meeting rooms, lounges, book-checkout areas, and a 120-seat lecture hall. The two subsurface floors contain book stacks, reading areas, and administrative offices. Opened in July 1978, the Fort Worth Central Library boasts an air-conditioning system 20 percent

smaller than that normally required in a comparably sized surface building. As a result, $3 to $5 million in energy costs will be trimmed over the
structure's forty-year design life.

BROOKLYN CHILDREN'S MUSEUM

The oldest institution of its kind in the United States, the Brooklyn
Children's Museum was founded in 1899. For seventy years, the facility
occupied two Victorian mansions in Brower Park, located in Brooklyn's
Crown Heights section. In 1969, these badly deteriorating mansions were
razed; the museum occupied temporary quarters until its $3.5-million
building was completed in 1977.

The Brooklyn Children's Museum is still located in Brower Park,
but instead of resting atop the park, the building is buried beneath it. A
turn-of-the-century subway kiosk is the entrance to a 180-foot-long "people tube" which cuts diagonally across the length of the main level. Stag-
gered on four tiers, this exhibition area offers children a "participatory
learning environment" complete with 20,000 items which children can
touch and explore for themselves. Exhibits cover natural history, technology, biology, and cultural history; a small oil tank located in one corner of
the facility serves as a museum theater.

The bottom floor of the two-story structure rests forty feet beneath
the surface. This floor is occupied by storage areas, darkrooms, work-
shops, and a resource center. A sunken courtyard emits daylight into one
corner of the exhibit area, and the 30,000-square-foot reinforced-concrete
building also features a skylight and clerestory windows. Wood floors and
ceiling beams blend with bright colors to produce a pleasant, airy environment.

Designed by Hardy Holzman Pfeiffer Associates, the Brooklyn Children's Museum offers as many attractions on its roof as it does inside. A
variety of urban sculptures dot the parklike landscape, including an ex-
pressway sign, a grain silo (which serves as the exit), and bleacher seats
built around an open-air theater. Because of the building's earth-sheltered design, children are able to enjoy a park as well as the museum.

9

Home Sweet
Earth-sheltered Home

In 1970, about ten earth-sheltered homes were scattered across the United States. Few of the owners of subterranean homes realized they were on the ground floor of a minirevolution in architecture. Since that time, the number of earth-sheltered homes has increased sharply. Although the precise figure is not known, most knowledgeable estimates place the number of underground homes in the United States between two and three thousand in 1980. Nearly everyone in the earth-shelter movement expects these figures to leap geometrically in the years to come.

Why such optimism?

Extraordinary public interest is certainly one reason. The University of Minnesota's Underground Space Center, a national clearing-house for information on earth-sheltered design, receives up to four hundred inquiries a day. No other research conducted at the university has generated such a response. Other campus clearing-houses are experiencing a similar flood of information requests. In the past few years, earth-shelter seminars, slide shows, and open-house tours have all drawn large, eager crowds. Owners of earth-sheltered homes are often hard pressed to maintain their privacy as curious strangers plead for quick "inspection tours."

Enthusiasm for underground homes is not spread equally across the country. There is a definite earth-shelter "hotbed" located in a wide band stretching from Minnesota to Texas. Although pockets of activity can be found in the Pacific Northwest, New England, and the Southeast, the majority of earth-sheltered homes are being constructed in the Midwest. Considering the weather, this is not surprising: the sweltering summers and freezing winters of the upper Plains are as notoriously brutal as the Siberian tundra. States such as Texas and Oklahoma have also buried homes within the earth for protection against storms.

Earth-sheltered homes are predominantly a rural phenomenon as well. With few exceptions, most homes are constructed far from large urban areas, unhindered by the constraints of zoning ordinances and building codes. Nor are rural earth-shelter efforts impeded by the pressures of architectural conformity so common in cities and suburbs. Burying a home underground in the middle of a dense city block may not violate local codes, but it is likely to raise the eyebrows—if not the wrath—of neighbors.

Because earth-sheltered homes represent such a stark break with conventional architecture, it is not surprising that many subterranean residences are do-it-yourself creations. These endeavors range from small one-bedroom homes to sprawling structures carved into the sides of gently rising hills. Not all have been unqualified successes. By their very nature, earth-sheltered homes require special attention to construction to prevent the building from leaking or sagging under the weight of the earth. Nor are such owner-designed homes likely to capture the imagination of home owners who can barely repair a gutter. Yet most of these underground do-it-yourself efforts are solid, well-built homes, apt tributes to the ingenuity and dedication which gave birth to their design.

Whether an earth-sheltered home is designed by an architect, a contractor, or the owner, it appears disarmingly normal when viewed from inside. In nearly every case, sunlight abounds. Windows in conventional homes frequently average 10 percent of the total wall surface, while in earth-sheltered homes the figure is often as high as 20 percent. The layout of the rooms, the building materials used, and the patterns of circulation within the home give little or no clue that the building is underground.

The exteriors of earth-sheltered homes run the gamut; like their above-ground counterparts, underground homes hold no monopoly on eye-pleasers or eyesores. Some are attractive, handsome structures, a welcome addition to any neighborhood. Others are downright ugly. Most fall in between.

Styles vary widely. A three-story earth-sheltered home in Colorado is nestled inside a sandstone cave, the living and bedroom areas exposed to plentiful sunlight. In Oklahoma, underground pioneer Jay Swayze has buried a chamber-style home completely underground. Residents and guests enter through a "beauty wall" which resembles a conventional suburban-style home and belies the windowless presence below. In suburban Minneapolis, a two-story 2,800-square-foot elevational home is so well-designed and attactive that it can easily withstand comparison with any surface dwelling. In Illinois, at least two earth-sheltered homes boast indoor swimming pools.

Despite such diversity, most earth-sheltered homes share at least one design trait in common: they lack a basement. A staple in many conventional homes, a basement is too costly to build in most earth-sheltered residences. For this reason, machinery and heating equipment are usually stored in a specially designed mechanical room located at the back of the home.

Earth-sheltered homes are no strangers to that other residential staple, the garage. Some are above-grade and attached to the entranceway; others are nestled into the earth alongside the home. In either case, the garage serves the same function as it does in any home: protection for the family auto as well as a handy storage area.

No matter what the style, location, or size, earth-sheltered homes are all big energy savers. Even if they were designed solely to preserve the landscape or provide storm protection, it is nearly impossible for an underground home not to provide a substantial reduction in utility costs. Some home owners are more energy conscious than others. They often employ such energy-saving devices as heat pumps, thermal window shutters, and Franklin stoves to reduce energy use. Occasionally, underground energy saving goes the extra mile. A few home owners have tapped wind power to reduce fuel needs. Others utilize water-filled glass jars inside stone columns to trap passive solar gain in winter. Some earth-sheltered residents prevent heat loss through windows by installing a system which shoots polystyrene beads between the panes of glass at night, removing them to storage the next morning.

Because earth-sheltered homes are such proven energy savers, many prospective buyers are undeterred by costs. The price of such homes varies widely, depending on such variables as labor, materials, and excavation requirements. Although some earth-sheltered homes are cheaper to build than their conventional counterparts, most range from 5 to 15 percent higher. This greater initial investment is ultimately recouped in substantially lower fuel bills and maintenance costs.

What follows is a brief look at a few of the best earth-sheltered homes in the United States. Each represents not only a different geographic location, but also a unique use of the earth-sheltered design. Each is eminently qualified to serve as a model for future underground residential construction.

SUNCAVE

Much of New Mexico, especially the northern part of the state, is anything but balmy in the winter: snowstorms and frigid temperatures are commonplace, as residents can well attest. New Mexico does have a de-

Completed in 1979, this large two-story earth-sheltered home in Arden Hills, Minnesota sold for $165,000. (Photo: David Martindale.)

cided advantage over many other states, however; it offers abundant sun light, especially in the winter. It was precisely this solar energy which architect David Wright sought to tap when he designed the Suncave near Santa Fe.

An expert in passive solar design, Wright is part of a small architectural firm called SEAgroup, located in Nevada City, California. Dedicated to designing energy-conserving buildings, each SEAgroup architect takes into consideration the following three factors when designing a home: (1) the needs and desires of the client, (2) the specific site characteristics; and (3) the local climate. What results is a unique solution to every design problem. The Suncave is no exception.

Located on five acres of wooded mountain land, this 1,385-square-foot home is a rich blend of three high-mass building materials. The walls are constructed of concrete and adobe, the latter being a common ingredient of the area's traditional pueblo homes. The floors are made of brick. Walls on the north and the northeast are nestled snugly into the earth, and most of the roof is covered with sod. A glass wall and companion clerestory face true south, welcoming the New Mexico sun.

Building codes and lending agencies would ordinarily require such a home to have a conventional heating system. However, because of the undisputed energy savings resulting from Wright's passive solar design, the Suncave lacks even a back-up furnace. Interior temperatures remain between sixty-five and seventy-five degrees F. year round. The adobe fireplaces provide what little additional heat is needed; during a typical winter, the occupants consume less than one cord of wood. Overhangs block sunlight in the summer and keep the home pleasantly cool.

Completed in March 1977, the interior of the four-room home is a visual playground of rich earth colors and hand-crafted wood details. Pineboard roofs slope down over the living area, and skylights admit additional light over the dining room and kitchen. Two bedrooms and baths are situated toward the rear of the home, where the earth-bermed walls block the biting north winds which buffet the hillside.

EARTHTEC 5

The Winston House and Baldtop Dugout brought architect Don Metz considerable attention. Both homes are impressive for their energy efficiency, superior design, and craftsmanship. Encouraged by the enthusiastic acceptance of these homes, Metz is now working almost exclusively on designing other earth-sheltered homes. The Lyme, New Hampshire architect formed a corporation called Earthtec, and in 1979 he built a prototype of a mass-market home.

Called Earthtec 5, the 2,000-square-foot single-level residence is located beside a narrow mountain road in west-central New Hampshire. From the parking pad, only an arched entranceway and a skylight are vis-

The south wall of the Winston Home provides abundant sunlight and a spectacular view of the New Hampshire countryside. (Photo: David Martindale.)

Two parallel steel culverts form the earth-sheltered Clark-Nelson home in west central Wisconsin. (Photo: David Martindale.)

An aerial view of John Barnard's Ecology House shows the central atrium and the flat-plate solar collectors used to trim utility costs. (Courtesy: John Barnard.)

The University of Minnesota's Williamson Hall features an admissions and records area on the left and an underground bookstore on the right. (Courtesy: Myers & Bennett Architects/BRW.)

aldtop Dugout in New Hampshire is designed to take maximum advantage of passive solar energy. (Photo: David Martindale.)

Slanted windows beside an atrium admit light inside the admissions and records area of Williamson Hall. (Courtesy: Myers & Bennett Architects/BRW.)

Overhead wooden beams dominate the interior of earth-sheltered St. Benedict's Abbey. (Photo: David Martindale.)

Designed by David Wright, the Suncave home in Santa Fe, New Mexico features walls made of concrete and adobe. (Courtesy: David Wright.)

Pat Clark prepares dinner beneath an overhead skylight in the kitchen of the Clark-Nelson home. (Photo: David Martindale.)

Stone walls in the living room of the Davis Cave provide an abundance of rich texture.
(Photo: David Martindale.)

The Crowell House was designed as a vacation home by architect Mark Simon. (Photo:
David Martindale.)

collector panels rest atop the twelve-unit Seward West townhouse project in Min-
polis. (Photo: David Martindale.)

two-story earth-sheltered home in Burnsville, Minnesota was built as part of an earth-
tered demonstration project. (Photo: David Martindale.)

A fully loaded truck leaves the International Trade Center located in limestone excavations beneath Kansas City, Missouri. (Courtesy: Great Midwest Corporation Subterropolis.)

In the future, the landscape will be dotted with skylights as more and more earth-sheltered homes are constructed. (Photo: David Martindale.)

Forest House, central Florida. (Courtesy: William Morgan Architects.)

ible. Upon walking inside, the visitor finds an attractive three-bedroom home laid out in a conventional rectangular pattern. Ceilings consist of exposed timber with spruce decking. Only the utility, storage, and mechanical rooms are devoid of direct sunlight.

A total of 330 square feet of triple-glazed glass form the south and west walls of Earthtec 5. According to Metz, winter sunlight streaming through these windows accounts for nearly half the energy needed to heat the home. The passive solar design provides a built-in mass of eighty tons which effectively stores the heat for later slow redistribution. Additional heat is supplied by a small conventional furnace; Metz estimates the home will use less than 600 gallons of heating oil a year. An optional wood parlor stove provides additional warmth. During the summer, tall trees located in front of the windows effectively prevent the sun's rays from warming the interior of the home.

A wood-burning stove inside Earthtec 5 takes the chill off the home on cold winter mornings. (Photo: David Martindale.)

DUNE HOMES

Walk along the ocean in Atlantic Beach, Florida and you are likely to see two strange-looking portholelike windows peering out from a sand dune. Built in 1975, these futuristic, mirror-image apartments were designed by noted Jacksonville, Florida architect William Morgan. Determined to preserve the serenity of the existing dunes and not obstruct neighbors' views, Morgan chose to bury his egg-shaped creation.

The Dune Homes were built by first pouring the concrete floors and then erecting a 2.5-inch-thick concrete shell which was designed with the aid of a computer. What resulted is an extremely strong bubble, easily capable of supporting the 20-inch-thick earth cover.

Upon entering either of the dune apartments, the visitor walks onto an overhanging balcony where a bedroom and bath are located. A curved staircase descends to the ultramodern living area below. A wood-paneled dining and kitchen section is situated directly beneath the bedroom.

Built at a cost of $50,000, the Dune Homes use about half the electricity as comparably sized surface dwellings. Although there is some heat gain early in the morning through the east-facing windows, the indoor temperature usually remains at a near-constant seventy degrees F.

DAVIS CAVE

In the early 1970s, builder Andy Davis received a monthly gas bill of $167 for heating his rented home in central Illinois. Andy Davis was not pleased. After pondering what to do, Davis announced that the family would declare war on soaring fuel costs by building their own home. And not just any home—a cave!

Together with one of his sons and a son-in-law, Davis built an eight-room, 2,000-square-foot underground home in Armington, Illinois. The main living area is octagonal, the bedrooms fanning out from a central living area. A rectangular annex with additional bedrooms and living area is attached to the main building.

The most unique feature of the Davis Cave is that it looks like a cave. From the outside, large stones protrude from the thick walls. Two huge oval windows admit sunlight to the living area but not to the bedrooms. Inside the home, shining stones of various sizes are embedded in the walls and ceilings, providing a limitless array of texture and an aura of strength to the structure. To sit inside Davis Cave is comparable to being inside a well-protected fortress. More than three feet of earth cover the cast-in-place reinforced-concrete roof, giving the home a solid, secure feel unmatched in surface dwellings.

Davis Cave is well equipped to withstand the brutal onslaught of Illinois winter. Most of the heat is provided by a Franklin stove; a fan system evenly distributes the air to every room. During the winter of 1978–1979,

a massive storm descended upon Armington, and Davis was without firewood. For several days, the family remained snowed in, virtual prisoners
in their own cave. Even unheated, the temperature inside Davis Cave
dropped just two degrees F. each day until it finally stabilized at sixty-two
degrees F. That winter, Davis spent a total of $1.29 to heat his home—the
cost of the gasoline needed to power the chain saw which he used to cut
kindling wood. During an average winter, Davis requires just 2½ cords of
wood to remain warm. During summer, temperatures inside the home remain about seventy degrees F.

Convinced he had developed a truly energy-efficient alternative to
the conventional home, Andy Davis began franchising his concept in
April of 1977. By 1980, he had sold more than 100 Davis Cave franchises
in various parts of the country. At least seventy similar cavestyle earth-
sheltered homes have been constructed nationwide. According to Davis,
most can be built for 5 to 10 percent less than surface structures. Not all
home owners prefer the stone walls; some opt for paneling, dry wall, or
stucco. Nor are prospective buyers limited to a standard-sized structure.
A doctor in Bloomington, Illinois built a $200,000 fourteen-room cave
complete with swimming pool, three-car garage, and office space.

Although the word *cave* may evoke less than pleasant associations,
Davis insists it is not only an apt description of his creation but also a definite attention-getter for these unique residential energy savers.

CROWELL HOUSE

Picture the following scene: a large open hill in central Vermont
with a commanding vista of the Green Mountains forty miles in the distance. During the spring and summer, the area takes on the appearance of
an undulating green carpet. Autumn is as though a celestial artist had
suddenly spilled bright splashes of yellow and deep red across the landscape.

Now suppose you wanted to build a home on this hill, but you
wanted it to blend in with rather than mar the site. Where would you put
it? You could do what architect Mark Simon did: slice the home into the
side of a hill.

Simon, who practices with Moore Grover Harper in Essex, Connecticut, was asked by relatives to design a vacation home on their fifteen-acre
treeless plot in Vermont. In order to keep the home from resembling
what Simon called "a German gun emplacement," he decided on the
earth-sheltered option.

Crowell House's clapboard façade makes it look like an ordinary
home. It's not. An eighteen-inch earth cover rests on the roof of the concrete structure, with insulation wedged between the earth and walls.
Coal-tar waterproofing provides an effective barrier against moisture. The
600-square-foot cottage has one master bedroom, a bath, a kitchen, and a

The original Davis Cave in Armington, Illinois was designed and built by Andy Davis. (Photo: David Martindale.)

living area, all heated solely by two wood stoves. Four skylights provide additional light and good cross-ventilation. The front porch shades the home in the summer. Because of the passive solar design, the house stays cool and dry in the summer and is virtually energy independent.

MINNESOTA DEMONSTRATION HOMES

When construction was completed on the twelve-unit Seward West townhouse project near downtown Minneapolis, these two- and three-bedroom homes were opened to the public for inspection. The resulting crush of curious visitors surprised nearly everyone. As many as 500 people daily would inspect the demonstration model. By the end of the ten-week open-house period, more than 10,000 Minnesota residents had visited Seward West.

Were Seward West just another townhouse, it would not have attracted such huge crowds. Seward West is the first earth-sheltered townhouse, and one of the first urban underground residential projects built in the United States. Located adjacent to I-94, a busy interstate highway, the back of the twelve-inch-thick concrete structure is bermed with more than a foot of earth. This soil cover not only trims energy costs, but also acts as a barrier against the din of traffic. The south-facing front façade welcomes sunlight with large windows and active solar-collector panels on the roof.

Each of the two-story homes at Seward West receives approximately 30 percent of its annual heating from passive solar energy. Another 45 percent of the heating load is supplied by retaining the heat from the active collectors in four-ton rock storage areas beneath each unit. Once the two-day supply of storage heat is exhausted, a conventional gas furnace automatically trips on. The builder, Seward West Redesign, Inc., estimates that the annual heating bill will total just $47.

Constructed in 1979, the story of Seward West began in 1977. Minnesota legislators, encouraged by the energy-saving features of earth-sheltered designs, appropriated $490,000 to the Minnesota Housing Finance Agency (MHFA) for the design and construction of eight earth-sheltered projects: five single-family dwellings and three park ranger stations. Single-family homes were required to integrate passive and/or active solar features with the earth-sheltered design. Under the proposal, private builders and developers were guaranteed $17,000 to cover any unusual costs in building and designing these unique homes. In return, the builder and developer agreed to select the site, obtain conventional financial backing, and sell the homes on the open market.

As the Minnesota legislature clearly indicated, the purpose of the grant was threefold: (1) to provide an opportunity for the public to examine earth-sheltered homes firsthand; (2) to test the market value and public acceptance of the earth-shelter option; and (3) to monitor the homes' construction and operating costs and energy efficiency. As part of the grant, the Underground Space Center was awarded $140,000 to monitor the eight demonstration homes and two control homes. The latter are conventional, all-electric, single-family homes. Sensing equipment for all buildings consists of an array of temperature moderators. The experiment is expected to last two years.

The grant to MHFA also resulted in four detached earth-sheltered homes throughout Minnesota. A typical example is a 2,000-square-foot home tucked into a ridge in the Minneapolis suburb of Burnsville. It was completed in April 1979; the first day of open house—a rainy, gray Sunday afternoon—saw more than 1,500 people tour the home in three hours. Many people saw the long lines and left without getting out of their cars. Others had to be turned away.

Built by Carmody Ellison builders, the large two-story home is completely topped with earth on the roof and bermed on 60 percent of the side walls. Large windows permit ample sunlight to flood the home during winter. At night, tenants close the thermal drapes to lessen heat loss. Boasting a wood-burning fireplace in the living room, the attractively designed Burnsville home is expected to run up a heating bill of only $100 per season, compared to bills of $200 or more *per month* experienced by many Minneapolis–St. Paul home owners.

Understandably, earth-shelter zealots across the country are keeping a close eye on these Minnesota demonstration homes. Most are convinced the experiment will prove that well-designed earth-sheltered homes are not only energy savers, but also viable entries in the residential marketplace.

10
To Live Beneath
the Soil

"Since I've moved here," says earth-shelter convert Dr. Margaret O'Connor, "being enclosed by the earth has taken on a very personal meaning."

What is it like to live beneath the soil? Reactions vary from one owner of an earth-sheltered home to another. Nearly all have had to make some adjustments, however minor, to their new environment. We were all born and reared in conventional surface dwellings. For this reason, the decision to live in an earth-sheltered home represents a radical break with residential tradition.

What follows are brief profiles of three owners of earth-sheltered homes. Seen through their eyes, the earth-sheltered home becomes a personal statement, mirroring the individual who lives there.

MARGARET O'CONNOR

The first time Margaret O'Connor noticed the Seward West townhouse project, she was driving to work along the freeway and spotted construction activity beside the road. At the time, the young Minneapolis physician was an earth-shelter neophyte. She had read a bit about the subject in the popular press, but the idea of living underground seemed more of a curiosity to her than an option.

Precisely because she was curious, O'Connor decided to inspect the Seward West demonstration unit in the fall of 1979. She liked what she saw. She was impressed by both the design and the energy-saving potential of the home. A few months passed before she decided to buy a two-bedroom unit. In February 1980, Margaret O'Connor became one of Seward West's first occupants.

A native of the Minneapolis area, O'Connor grew up in a home in which her bedroom was located in the basement. As a child she realized that despite soaring summer temperatures, her room stayed pleasantly cool. More recently O'Connor lived in an old uninsulated duplex in Minneapolis. Despite the fact that indoor temperatures in the winter seldom went above sixty degrees F., she received monthly heating bills as high as $70.

For Margaret O'Connor, the days of expensive heating bills are now over. Even before the heat was turned on in her new home in late January, the indoor temperature registered fifty-five degrees F. During the first fifty-five days she lived in her townhouse, her heating bill totaled $37. The fact that Seward West receives the bulk of its heat from the active and passive solar design has even altered O'Connor's perception of the world around her. "I'm very much more aware of the sun now," says O'Connor, pointing to the large south-facing windows. "And I'm very unaware of temperature."

O'Connor is deluged with a variety of questions when people learn where she lives. The most typical is, how do you deal with the dark? Although she is tempted to smile, O'Connor patiently explains that she receives ample sunlight throughout her home. And once friends or relatives visit her at Seward West, they are rather incredulous that it is as light and airy as it is.

According to O'Connor, one of Seward West's biggest assets is its tranquility. The home is amazingly quiet, and she hears hardly any of the traffic roaring behind her home. In fact, O'Connor views her townhouse as a large nest, one which evokes a very secure, safe feeling. The contrast with the world outside can be stark, though; once she walks out the door of her earth-sheltered home, she is in the heart of a major city just a mile and a half from downtown Minneapolis.

O'Connor realizes she is a pioneer of sorts, although she views her decision to live at Seward West as practical rather than radical. She is a genuine conservationist: her home consumes far less energy than most in America. During the summers, she grows fruits and vegetables in the front yard, a bit of self-sufficiency which also saves energy. Since the hospital where she works is just three blocks from Seward West, her car often sits in the garage for days at a time. She can go a month without visiting a gas station. Should energy costs take even greater leaps in the future, Margaret O'Connor will be counted as one of the survivors.

Because the back of the Seward West townhouse project is earth-bermed, residents barely hear the street traffic from a nearby freeway. (Photo: David Martindale.)

BUDDY WRIGHT

When neighbors first saw Buddy Wright's new home being buried in the ground near Hobart, Oklahoma, they told him he had a lot of guts. Wright's three-bedroom earth-sheltered home resembles a big doughnut. It raises as many eyebrows today as it did when it was built.

Although there are at least half a dozen earth-sheltered homes within a fifty mile radius of Hobart, Wright's attracts the lion's share of curiosity. During the first year and a half he lived there, more than three thousand people visited the home. According to Wright, many arrived with a dungeonlike image of what it was like to live underground. This image quickly vanished upon touring Wright's creation.

Avoiding rectangles and squares, the Wright residence is a circular home fifty feet in diameter. A twenty-foot solarium sits in the center of the structure, topped by a geodesic dome, the only object visible from the roadway. Thirty feet of the south wall is exposed, providing both an entranceway and additional glazing. These windows offer a view across a creek toward some small mountains, and also permit the home to take advantage of passive solar gain in the winter. Although the home is completely below grade, so much sunlight penetrates every room that Wright insists his home is brighter than most above-grade structures.

Buddy Wright first became interested in earth-sheltered homes in 1975. "I looked at the energy situation and in my own mind, I determined that it was going to get worse," says the automotive parts dealer. "So I decided to do something about it." Soon afterward, he read about a geodesic-domed earth-sheltered home built by engineer Paul Isaacson in Utah. Intrigued by the concept, Wright designed a similar underground home of his own. Wright had already constructed a conventional surface structure by himself, so he decided to do the same thing with this more unique design. Together with a friend, Wright completed the 2,000-square-foot residence in six months. He and his wife moved into the completed building on February 1, 1979.

Buddy Wright estimates he saves approximately fifty percent on his energy costs thanks to the earth-sheltered design. Most of his heat comes from burning logs in the fireplace. A back-up heat pump is available for additional warmth. In the summer, the home stays comfortably cool despite outdoor temperatures in the 90s. Only when the temperature remains in the 100-degree range for several days do indoor temperatures climb to 80 degrees. An exhaust fan in the solarium not only removes heat build-up during summer, but also provides good cross-ventilation.

A practical man by nature, Wright insists that living underground is just about what he thought it would be like. "It's very, very comfortable," says Wright. "You don't feel any different in this home than you would in a conventional home." In addition, there is the advantage of knowing that the home is virtually fireproof, stormproof, and maintenance-free. A combination fiber glass–urethane-foam moisture sealant has kept the house dry. "We haven't seen a single drop of moisture," says Wright.

Some owners of earth-sheltered homes have difficulty adjusting to the new-found quiet once they begin living underground. Not the Wrights, who have lived in the country and are used to the silence. The only thing the Wright weren't used to was all the attention. Buddy Wright doesn't think the curious will stop gawking at his home anytime in the near future. He knows that he has built a most unusual home, one that is as unique as it is practical. Until other home owners in Oklahoma start building circular earth-sheltered homes with geodesic domes, the questions and the stares are likely to continue.

TOM AND MARY HAGAN

In August 1979, the Hagans and their two children moved into a new two-story earth-sheltered home outside Preston, Minnesota. A few months later, they decided to hold an open house and let the community inspect the home. Announcements appeared in the local newspapers, noting that the Hagan home would be open between 1 P.M. and 5 P.M. on a

Saturday afternoon. Tom Hagan says he realized at the time that earth-sheltered homes generate a great deal of enthusiasm, but he was dumbfounded at the crowds which showed up at his door.

The Hagans are uncertain exactly how many people visited the home. "We stopped counting at two thousand," says Tom. Some people drove to the rural southeast Minnesota location from as far away as South Dakota. Traffic was so heavy that the sheriff had to be called out to unsnarl the tie-ups.

Despite the crush, the Hagan open house was a success. Tom and Mary had succeeded in raising visitors' consciousness about the earth-shelter prospect. According to Tom, most of those who visited the home expected to view a cave. Instead, they came away delightfully surprised. A typical reaction: "This just like any other house—only it's beautiful!"

Located on ten acres of heavily wooded land, the Hagan home is an elevational design set into a gently rising hill. All the windows face southeast. Bedrooms are located upstairs, and the kitchen, dining room, study, and utility room are situated below. The living room is also on the first floor, open at the top and surrounded by a railing on the second level.

The Hagan home was built primarily in response to the energy situation. Before they moved into their new home, the Hagans lived in a large old frame house which each winter devoured $150 worth of heat per month. Tom's concern about energy costs was heightened when he read an article in the *Wall Street Journal* which predicted a 300 percent increase in natural gas prices. About the same time, Tom Hagan, a chiropractor, eyed an article which described an earth-sheltered home in the Twin Cities area. He quickly put two and two together and realized the second article suggested a solution to the problems posed by the first. For this reason, the Hagans decided to nestle their home into a hill and become virtually energy independent.

Because they built in a rural location, the Hagans encountered no restrictive building codes or zoning ordinances. Nor did they have any trouble financing their home, although Mary concedes that they were "rather lucky" to find a progressive lender. They did, however, have problems with their contractor. Unfamiliar with earth-sheltered construction, the contractor decided to err on the side of caution. The back concrete walls of the Hagan home are twenty inches thick, a bit of structural overkill as expensive as it is unnecessary. What was supposed to have been a modest two-bedroom home turned into an 1,800-square-foot structure which was much larger and more expensive than the Hagans had planned. "Our builder simply got carried away," explains Tom. "As a result, we got a lot more house than we figured."

The house the Hagans did get is extraordinarily energy efficient. Because the home utilizes both active solar panels and passive solar design, it doesn't need a conventional back-up mechanical system. There is no need for air conditioning during summers, since the home stays cool even on the warmest days. During winters, additional warmth is provided by a circular fireplace, the forest providing ample firewood.

Sometimes the home's thermal efficiency amazes even its owners. During Christmas of 1979, the family left the home unoccupied for five days while visiting relatives. When they returned, the indoor temperature was a comfortable sixty degrees F., despite the fact that Minnesota had been experiencing a prolonged period of subzero temperatures. "We couldn't believe it," says Mary. Some day the Hagans hope to increase their energy self-sufficiency by installing a wind generator to provide electricity.

Except for its size, the Hagans have few complaints about their earth-sheltered home. Mary relishes the silence the building affords, although Tom admits it is sometimes so quiet that he is awakened by the sound of a woodpecker outside the bedroom window. Tom and Mary are avid vegetarians, and their rooftop garden flourishes with a wide variety of home-grown produce. The home also boasts a root cellar, similar to those so popular in pioneer days.

"I like this home because it's the most harmonious with nature," says Mary. "It takes so little and yet offers so much." The home also affords the Hagans the opportunity to live relatively independent of corporate energy sources. "I don't really think of ourselves as energy pioneers," says Mary. "I just look at this home as a way to keep ourselves free."

11
Cornering Caverns
for Commerce

In the excitement generated by earth-sheltered architecture, we sometimes lose sight of that lowliest of subsurface habitats, the cave. But caves are making a comeback: man-made caves are being tapped for commercial and other nonresidential purposes. Developers are finding advantages similar to those of earth-sheltered buildings, not the least of which is the extraordinary savings in energy.

Who would blast a hole in the earth to create a cave? The military, for one. The Strategic Air Command headquarters sits safely underground beneath Omaha. Missile silos and support facilities are buried at various locations throughout the country. In 1965, the North American Air Defense Command (NORAD) moved into its new home in Cheyenne Mountain, outside Colorado Springs. This sprawling computerized listening post consists of eleven buildings, some as tall as three stories. Each sits on springs and shock absorbers designed to cushion the shock from blast waves. NORAD's submerged hideaway could survive any nuclear volley short of a direct hit.

The purpose of subgrade military installations is protection, and underground protection is the concern of private firms, too. Consider the Iron Mountain Security Storage Corporation (IMSSC) in Hudson, New York.

Formed in 1951, IMSSC provided a combination survival shelter and alternate headquarters for some of the country's corporate giants. In the event of a disaster (nuclear or otherwise), executives of such firms as

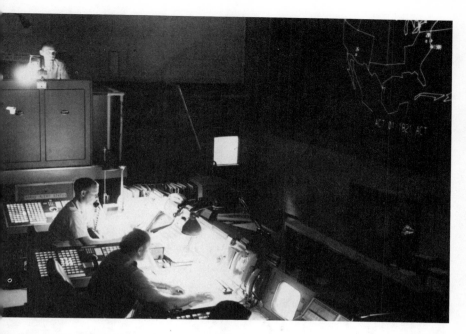

The headquarters of the North American Air Defense Command is buried deep inside Colorado's Cheyenne Mountain. (Courtesy: NORAD Public Affairs.)

Shell, Standard Oil, and Manufacturers Hanover Trust could flee to this former iron mine and find safety beneath 75 to 150 feet of solid rock. Tenants were guaranteed thirty days of emergency power, water, plumbing, and radiation-scrubbed air.

Each disaster headquarter was built to order. The area leased by Standard Oil was designed to hold two hundred executives and their families in fifty-nine separate rooms. Painted in bold colors, the Standard Oil encampment provided desks, meeting rooms, and even piped-in music.

By the mid-1960s, however, the demand for such bombproof alternative headquarters evaporated. For this reason, the Boston-based Iron Mountain Group converted the Hudson, New York mine into a storage facility for vital records. Two other similar underground record-storage areas are located in an abandoned limestone cavern in Rosendale, New York and an earth-covered bunker near Providence, Rhode Island. Approximately two hundred of the Fortune 500 firms store their vital records with the Iron Mountain Group. Many even maintain complete computer libraries in these well-protected underground facilities.

Hundreds of miles away, in Hutchinson, Kansas, a sixty-year-old salt mine performs a similar function. First opened in 1960, Underground Vaults and Storage Company (UVS) is located 650 feet below ground in a 40-by-100-mile salt mine which is still being excavated. In one corner of this cavity, UVS guards the most sensitive records of more than fifteen

Salt mines like this one in Hutchinson, Kansas offer near-perfect conditions for storing vital records. (Courtesy: Underground Vaults and Storage, Inc.)

thousand firms and individuals. These data include computer tapes, microfilms, and millions of paper documents. Larger firms lease concrete vaults, some of which house such items as hospital records, secret oil-company drilling data, and a complete library of Metro-Goldwyn-Mayer films.

Surrounded by salt deposits millions of years old, these records are protected by the constant sixty-eight-degree temperature and 50 percent humidity. Rooms are dry and free from rodents. Because of the unique location, security is virtually impossible to penetrate. Entrance to the mine is via a single elevator shaft in an isolated part of town. A few well-trained security personnel, several closed-circuit television cameras, and steel vaults are more than adequate to discourage even the most determined intruder. Because upkeep is minimal, the cost of storing records below ground trims rental fees by as much as 97 percent.

Iron Mountain and UVS are not the first groups to exploit the potentials of excavated caverns. In Poland, several salt mines host hospitals which treat patients with nervous disorders. Ever since 1937 an abandoned limestone mine near West Winfield, Pennsylvania has supported a mushroom-growing plant. A similar mine in Wampum, Pennsylvania houses the technical division of a large cement manufacturer. And deep within abandoned lead and zinc mines in Idaho, an experimental project

Underground Vaults and Storage in Hutchinson, Kansas provides ideal storage conditions for magnetic tape and other computer-related materials. (Courtesy: Underground Vaults and Storage, Inc.)

tested the feasibility of growing containerized trees underground for a fraction of what they cost in conventional surface greenhouses.

By far, the cavern capital of the nation—perhaps of the world—is greater Kansas City. Straddling the Missouri River, the Kansas City region sits atop a massive formation of 270-million-year-old limestone. Since mining began sixty years ago, more than 130 million square feet of rock have been removed in thirty sites. More remains to be extracted; at an excavation rate of 6 million square feet annually, mining operations are likely to continue through the year 2015.

Once limestone is gouged from beneath Kansas City, the mines are not abandoned; instead, private developers transform these vast caverns into sprawling underground industrial parks. By 1980, more than 20 million square feet of mined space in the Kansas City area had been converted to secondary use, mostly for storage and warehousing.

Although the secondary-space concept was first implemented in 1954, it wasn't until Amber Brunson's successful underground experience that shifting operations to limestone caverns became popular. Shrewd, opinionated, and a self-avowed business rebel, Brunson is founder and president of Brunson Instrument Company, a firm which manufactures high-precision optical equipment used on spacecraft and submarines. In the mid-1950s, Brunson was confronted with a critical problem: his factory shook. The constant rumble of traffic outside his downtown Kansas City plant produced vibrations which ricocheted throughout the building. After a while, such vibrations began to seriously hamper efforts to fine-tune instruments within specifications as rigid as fifty-millionths of an inch.

Determined to find a solution, Brunson recalled that underground factories were built in Germany and Sweden during World War II. He eyed the limestone quarries at the edge of the city, checked with a few geologists, and decided to carve a new plant underground. The entire structure was blasted out of rock according to Brunson's own specifications. Since the excavated limestone was crushed and sold on the open market, Brunson's plant cost one-third what a similar factory would have cost on grade.

In March 1960, Brunson and his firm said good-by to their shaky downtown location and moved into their rock-hewn factory. It does not boast a fancy façade; no imposing structure can be viewed from the interstate, and the entrance is barely visible from a distance. However, because it is buried seventy-seven feet beneath solid rock, the 140,000-square-foot plant is vibration free. As a result of his decision to relocate underground, Amber Brunson became a local celebrity and a well-known figure among underground enthusiasts.

The fact that a highly respected firm such as Brunson Instrument would set up shop in a limestone cavern had a definite impact on other businesses in the Kansas City area. Few were eager to blast away limestone to carve their own factories, but gradually firms chose to lease secondary space from the growing number of cavern developers.

As a result, mine operators altered their excavation techniques. Before secondary-space use caught on, roof-support pillars were randomly spaced and irregularly shaped, which was practical when the mines were to be abandoned after removing the limestone. But it didn't sit very well with prospective tenants who needed to set up long assembly lines or arrow-straight storage platforms.

When it became apparent that the excavated space was at least as valuable as the rock, mine operators and lessors worked out a solution. All quarrying in the Kansas City area is now done with secondary-space use in mind. Pillars are rectangular, aligned in a row, and evenly spaced at approximately forty-foot intervals. The result is a standard grid pattern of pillars, with ceilings twelve to thirteen feet high. To ensure safety, carbon dioxide readings are measured daily, and the rock is tested regularly for structural defects.

Unlike many mines, the limestone quarries in Kansas City are entered by tunnels cut into the cliffs at grade level. Except for the absence of daylight, the activity inside is similar to that of any conventional industrial park. Trucks stream through the tunnels, entering through brightly lit paved roads, and head for rock-protected loading docks. Diesel locomotives back boxcars into underground railroad sidings for loading and unloading. At the foreign trade zone, United States customs inspectors oversee the import and export of overseas goods. Throughout the caverns, stock clerks and secretaries, typists and truckers scurry about their business oblivious to the novelty of their location.

What entices Kansas City commercial concerns to relocate underground? A whopping 70 to 100 percent savings in energy is a key factor. At depths from 50 to 200 feet, the limestone caverns maintain a near constant temperature of sixty degrees. Humidity remains stable at 50 percent, and the entire area is completely unaffected by outside extremes of temperature. Lights, machinery, and people provide what little extra heat is necessary. If it gets too warm, tenants can pipe in cooler air from farther back in the caverns. Even when air conditioning is needed to further reduce humidity, the size of the system is greatly reduced. At Brunson Instrument, air-conditioning units are just one-tenth the size needed to cool a similar plant on the surface.

The largest lessor of secondary space is the Inland Storage Company, located under the Kansas side of the Missouri River. Besides conventional warehousing facilities, Inland boasts more than 24 million cubic feet of food-storage area. Some $300 million in food is frozen at minus five degrees F. Once a limestone locker has served as a freezer for a year, the rock surrounding the freezer becomes so cold that refrigeration needs are cut in half. Dr. Truman Stauffer, professor of geoscience at the University of Missouri at Kansas City, estimates that the energy saved by

Employees at the underground Inland Distribution Center in Kansas City, Kansas can unload eighty boxcars each shift. (Courtesy: Inland Storage Distribution Center.)

freezing food below ground is equivalent to the amount of energy used annually by 7,600 homes.

In the unlikely event of a massive power outage, Inland would not be worried about food spoilage. Even without electricity, freezer temperatures would rise at a rate of just one degree per day, rather than one degree per hour in comparable surface freezers. There's no need for costly stand-by refrigeration equipment; the back-up unit is the rock itself.

There are a number of other advantages to occupying excavated quarries. Rental rates are less than comparable fees on the surface, sometimes by as much as 40 percent. Because limestone caverns are fireproof, insurance costs are minimal. Expensive maintenance problems are virtually eliminated, and floors are able to withstand even the heaviest loads. Noise, dust, and vibrations are all easily controlled, and the submerged location lends itself to tight security.

And Kansas City's limestone caverns do far more than provide a one-shot economic boost to the area; because the limestone quarries aren't abandoned, they provide both a badly needed tax base and a long-term economic stimulus to the region. Secondary-space use translates into two thousand jobs, more employment than it took to remove the rock. "What was dead property after mining out the rock is now a source of steady income," a regional mine superintendant told Dr. Stauffer. "The quarry at 95th Street was nothing but rock, but now it's gold!" Because

The office of Waechtersbach USA, Inc. in the foreign trade zone beneath Kansas City, Missouri. (Courtesy: Great Midwest Corporation Subterropolis.)

Missouri has conserved its caverns, the state leads the nation in the secondary use of mined space.

Considering that limestone mining is producing space at ten times the rate it is being occupied for secondary use, there is certainly no shortage of space. But even though limestone caverns offer so many attractive assets, the developers of underground Kansas City are aware their location is also their biggest drawback. "There are some individuals who are just psychologically opposed to going underground," says Cherie White, manager of advertising and public relations for Great Midwest, Inc., a large lessor. "I call it the 'Tom Sawyer-lost-in-a-cave' syndrome. They automatically assume it's cold, dark, and wet in here."

Many prospective tenants change their minds when they visit the site and inspect the facilities. But often, enticing business planners to enter a quarry is nearly impossible.

The employees who work in Kansas City quarries do not regard their location as unnatural. They do park their cars beneath solid limestone, and they can't watch the weather change. But for the most part, working conditions beneath the cliffs don't differ much from those above them. There is no indication that employees are adversely affected by their location. To make sure, scientists are checking. Under the sponsorship of the National Academy of Sciences, Dr. Stauffer is conducting a study of the psychological effects of working in underground caverns.

The fifteen firms which lease secondary space in Kansas City have formed a nonprofit corporation called the Underground Developers Association. Their purpose is to share technical information and promote the use of underground space. Pride in their unique location seems to outweigh corporate rivalry.

"We don't consider each other to be the competition," explains association president Donald Woodward. "Our competition is on the surface."

12
Beneath the Far
Corners of the Earth

The use of underground space is by no means confined to the United States. Other nations also appreciate the underground option, and many utilize subsurface space for a variety of purposes. But despite soaring worldwide energy costs and ever-dwindling surface land, nearly all modern earth-sheltered homes have been built in this country. No other nation has matched the United States in constructing earth-sheltered buildings, residential as well as commercial.

At least part of the reason is climatic. Although approximately the same size as the United States, Europe is, for the most part, blessed with a far more temperate climate. Only in Scandinavia and the Soviet Union do winter temperatures plummet as they do in New England and the upper Midwest, and both of these former regions have embarked upon several ambitious underground projects. As for the Southern Hemisphere, few areas south of the equator experience great temperature swings. In equatorial regions where temperatures are high year-round, most nations lack the economic incentives to investigate the earth-sheltered design.

Such nations need not look only to the United States for earth-sheltered inspiration. Examples abound around the globe.

SWEDEN

For decades, the Swedes have been burrowing beneath their rocky landscape, carving out the bowels of their nation, tapping a resource they call "terraspace." Sweden has the necessary expertise, as mining has long been a backbone of Swedish industry. Not only is Sweden a leader in excavation technology, but this Scandinavian nation also exports this technology and equipment throughout the world.

Examples of Swedish subterranean achievements abound. A lack of suitable surface land in Stockholm was one reason Swedish planners decided to build three of the city's sewage-treatment plants underground, thus alleviating the problem of odors as well. Several Swedish factories are buried beneath the earth; one of these produces sensitive electronic equipment used in military hardware. The dustfree, vibrationless underground environment proves ideal for such precision manufacturing. Also located underground is a sizable percentage of hydroelectric power plants. Stockholm's archives are buried in a five-story rock-hewn structure.

Sweden began developing its subterranean resources on a large scale following World War II. Although neutral in its foreign policy, Sweden recognized the danger of an all-out European thermonuclear war. The government decided to protect its people and much of its military might by digging underground.

Nearly all large housing projects in Sweden are required to have air-raid shelters buried beneath them. Large public air-raid shelters are scattered throughout the country. The Katerina underground parking garage in Stockholm, which holds 550 automobiles, can be quickly converted to a civil-defense shelter capable of providing protection for twenty thousand people. Heavy steel and concrete doors swing shut, shielding occupants from radiation.

The most ambitious of Sweden's underground efforts are defense facilities. Nearly all Sweden's air force is based in underground hangars. In the event of an emergency, planes would be towed from their cavern hiding places to adjacent runaways. The giant Muskö naval base, located twenty-five miles south of Stockholm, is carved out of a mountain beside the sea. So vast is the interior of this unique base that navy destroyers are able to steam at full throttle into this rock-hewn fortress. Once inside, huge hydraulically operated iron doors seal the entranceway, protecting large drydocks, offices, and oil-storage tanks. Only a direct hit by a nuclear weapon could destroy this cavelike naval command center.

AUSTRALIA

Parts of the Australian outback are reminiscent of California during the bustling gold rush more than a century ago. The South Australian boomtown of Coober Pedy, located in the middle of a scorching desert 1,000 miles west of Sydney, exists because of the presence of a single valuable commodity: opals. The town lies near one of the world's

Portion of a kitchen in an underground home in opal-rich Coober Pedy, South Australia. (Courtesy: Australian Information Service.)

richest opal fields. Encouraged by a government policy which forbids large mining firms from extracting the gems, get-rich-quickers from more than forty nations descend upon this dusty town to make their fortunes. Most stay a year or two and then leave. Coober Pedy is one of the last areas in the world where the hard-working prospector can, with a little bit of luck, strike it rich.

In addition to its reputation as a hard-drinking, tough-talking frontier town, Coober Pedy is famous for its dugouts. Nearly half of the town's approximately four thousand residents live in these underground windowless homes carved into the soft clay stone. These homes are easy to construct and welcome respites from the scorching 100-degree F. temperatures which sear the town. Few of these dugouts are attractive; many are temporary structures which are abandoned once the owners move back to civilization. Occasionally, these dugouts bring more than respite from the heat: one dweller unearthed $900 in opals while digging a new bathroom.

Similar dugouts are common in White Cliffs, New South Wales, another opal-rich Australian community. Homes buried from seven to thirty-three feet beneath the earth take advantage of stable year-round temperatures of about sixty-five degrees F. Because the area lacks appreciable rainfall, home owners show little concern about moisture control. Skylights are used to brighten interiors, and at least some of White Cliff's electric energy is generated by windmills.

CANADA

Although Canadian winters are more severe than our own, relatively few Canadian home owners have availed themselves of the thermal advantages offered by earth-sheltered homes. Judging from Canadian attendance at earth-shelter conferences in the United States, it won't be long before Canada develops its own earth-sheltered building program.

There is one place in Canada where digging underground has long been a respected practice: Montreal. No stranger to brutal winters, this French-Canadian city has developed an elaborate network of underground tunnels and shops, the envy of any northern city. In 1962, a subterranean shopping center called Place Ville Marie opened to a less than enthusiastic reception. But once the snows arrived and temperatures dipped well below freezing, Montrealers made a very logical discovery: Place Ville Marie allowed them to shop and dine without bundling up against the winter cold.

In 1967, another huge underground complex, Place Bonaventure, was buried under Montreal. Linked with Ville Marie via subsurface walkways and the impressive metro, these complexes form a vast subterranean network which stretches over twenty-two acres. In recent years nearly every major commercial project in the area has some association with this winterized underground oasis. These subgrade shopping malls boast nearly seven hundred stores, as well as restaurants, theaters, movie houses, art galleries, skating rinks, banks, and indoor swimming pools. Thanks to this ambitious undertaking, winters in Montreal are now a bit more tolerable.

FRANCE

Established in 1946, the United Nations Educational, Scientific and Cultural Organization (UNESCO) was without a permanent home during the first few years of its existence. In 1952, UNESCO decided to build its headquarters in Paris. Architects proposed a Y-shaped high-rise bordered by a plaza. At first, Parisian authorities balked at the plan, claiming it was too tall and would mar the famed skyline. UNESCO eventually won approval, and the structure was built as planned.

By 1960, UNESCO headquarters was bursting at the seams, and officials decided to build an annex. When UNESCO presented its plan for a ten-story structure, the Parisian Commission des Sites rejected it.

What resulted was an earth-sheltered building buried thirty-five feet beneath the UNESCO plaza. Constructed around six large atriums, the three-story UNESCO annex provided 350 offices and a 400-car garage. Built to hide its presence, the structure is cited as an outstanding example of "invisible architecture." Many Parisians walk by the landscaped plaza never realizing that hundreds of UNESCO employees are busy working in offices buried below.

The French have also constructed an unusually shaped 1,000-seat underground amphitheater in Paris. Built beneath a courtyard, the structure is ovoid, in order to spread the weight of the building over a large surface. Resembling a huge concrete egg tilted at a slight angle, the amphitheater's roof gently rises above the courtyard, giving few clues to the nature of the structure below.

CHINA

The provinces of Shensi, Shansi, Kansu, and Honan are covered by loess, a volcanic soil. Loess is extremely fertile and very soft. Faced with the need to preserve agricultural land and house its people, China has been digging entire cities beneath this land since the 1920s. Today, more than 10 million Chinese live underground, perhaps the largest number of troglodytes ever to inhabit a single region.

Buried at depths of up to thirty feet, underground homes are built around courtyards. Entry is via L-shaped staircases, and the atrium-style design offers ample sunlight. Each home is protected from biting winds and temperature extremes. Most importantly, the buried cities offer China the opportunity to put the land to dual use. Each year, the soil is tilled, providing ample crops to feed the millions of inhabitants who live below.

For decades, American architects have been exporting the international style to every corner of the world. As a result, energy-inefficient buildings in Nairobi resemble those in Nashville. The new breed of earth-shelter architects are less doctrinaire in the export of American underground expertise. For this reason, future earth-sheltered construction worldwide will not always resemble similar buildings in the United States. But they will be highly energy efficient—a trait common to all earth-sheltered buildings regardless of their location.

(UNESCO's earth-sheltered annex in Paris is built around central atriums. Courtesy: UNESCO/Dominique Roger.)

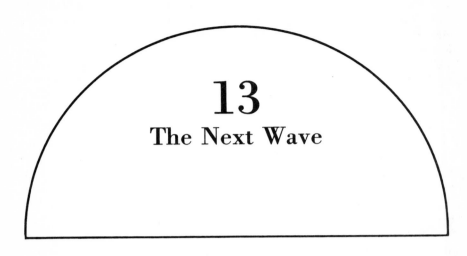

13
The Next Wave

Because of their impressive energy savings, earth-sheltered buildings are being built from one end of the United States to the other. Underground enthusiasts have difficulty keeping track of them all. University clearing-houses such as the Underground Space Center and periodicals such as *Earth Shelter Digest & Energy Report* catalogue as many as possible, though the ever-growing number of earth-sheltered buildings threatens to overwhelm their best efforts.

The larger, energy-progressive earth-sheltered projects receive deserved attention, and some residential projects—particularly mass-produced subdivisions—garner a great deal of publicity in both the technical and popular press. The impending next wave of earth-sheltered construction will probably raise as many eyebrows as the initial wave did in the 1970s. It may well prove instrumental in determining the long-term direction of the movement in the decades ahead.

This chapter is divided into two sections. The first examines the earth-sheltered buildings planned for the early 1980s and projects which might receive the go-ahead during the decade. The chapter concludes with a look at the prospects for mass-produced underground homes as viewed by the movement's three leading theoreticians.

In 1977, the state of California conducted a "Competition for Design of an Energy Efficient Office Building." Architectural and engineering firms from around the country were invited to compete, and a variety of creative proposals was submitted. After carefully reviewing the forty-one entries, the state announced the winner: a bold earth-sheltered design submitted by Benham-Blair and Associates.

Planned for a two-block site in downtown Sacramento, the Benham-Blair project features a six-story solar tower linked to a sprawling series of earth-sheltered offices. From a distance, the triangular-shaped solar tower will be the most visible feature of the 240,000-square-foot complex. The entire south-facing wall of the structure will be covered with 12,000 square feet of concentrator solar collectors. On sunny days, these panels will slowly alter position to track the sun as it passes above. A vertical courtyard behind the solar wall will provide light for the building's multitiered offices.

The solar tower will occupy the smaller site on one side of Sacramento's N Street. A subterranean commons will link the tower with a two-story earth-sheltered office building occupying a much larger city block. Divided into six modules, these subgrade offices will be built around landscaped courts open to the sky. Trees, benches, and sculptures will rest atop the parklike roof, and a meandering stream will traverse the entire length of a central mall.

Headed by architect Buford Duke, Benham-Blair's design team estimates that this project will save the taxpayers of California approximately 50 percent on energy costs. One reason is that the building will be responsive to Sacramento's climate. During most of the year, the state capital experiences relatively hot days and much cooler nights. The cool night air will be used to flush out unwanted heat built up during the day. Night air will also help chill water stored in tanks.

During daytime hours, the solar panels will heat water to about 350 degrees F. An absorption machine will convert the steam to chilled water to be used for cooling on warm days. When the solar panels can no longer supply additional cooling, the chilled water will be pressed into service. During winter, solar panels will provide forced-air heating; heat from mechanical systems will also be recycled.

There are subtle energy savers built into the Benham-Blair design. Photoelectric cells will shut off electric lights when sunlight provides adequate illumination. Pedestrians will be encouraged to use conveniently located stairwells rather than energy-gulping elevators and escalators. Window blinds will be adjusted to reflect sunlight in summer and admit it during winter. Perhaps the forerunner of similar energy-efficient public buildings, the California State Office Building is scheduled for completion in 1981.

Seventy miles to the west, the city of San Francisco is constructing an earth-sheltered building along Howard Street. Called the Moscone Center, this massive, 650,000-square-foot convention center will rest in the middle of an underground lake. Sump pumps and bentonite waterproofing will keep the building dry, and an eight-foot-thick concrete floor will prevent it from floating. Although one level of the four-story structure will be visible from the street, the rest of the eleven-acre building will be concealed below grade. When completed, the Moscone Center will become the world's largest earth-sheltered building.

At Oklahoma State University, architects are drafting a plan for an earth-sheltered housing project for the Southwest. Designed to house workers who will construct the MX missile system, the project is envisioned as two "passive" cities—one a large community, the other a satellite facility—both of which will be totally energy independent. Because these temporary cities will be earth integrated, they will require no heating or cooling to remain comfortable year round.

In the upper Midwest, the University of Minnesota is building an earth-sheltered campus building to house the Civil and Mineral Engineering (C&ME) Department, the parent body of the Underground Space Center. Designed by Meyers and Bennett Architects/BRW, the firm which conceived Williamson Hall, the $15.8-million structure will be built 95 percent underground and feature an atrium, numerous skylights, and a descending courtyard adjacent to the entrance. The first three floors of the 150,000-square-foot building will house classrooms, offices, and laboratories. A thirty-foot shelf of limestone will separate two deeper floors occupied by the environment and mineral engineering laboratories and the Underground Space Center.

The C&ME building will achieve considerable energy savings, trimming 50 to 60 percent of what would be required on grade. Heat will be generated by active collector panels, passive solar gain, and internal mechanical sources. The C&ME building will also feature a unique "ice air-conditioning" system pioneered by former University of Minnesota professor Thomas Bligh. As part of this system, a series of copper pipes will be attached to the solar collectors. A solution inside these pipes will circulate, absorbing cold from the winter air; this will be used to form a huge block of ice deep within the building. During the summer, the ice will chill air, which will then be delivered by another system for cooling.

Like Williamson Hall, the C&ME building will be closely monitored for its energy performance. Approximately $1.7 million in thermal-sensing equipment will be built into the design, and a variety of federal and private funds will be used to conduct an array of energy experiments. The building will also employ a light and prism system whereby sunlight will be "piped" into those areas of the structure which are unable to receive direct daylight.

Other Myers and Bennett earth-sheltered projects either under construction or planned include a student center at the St. Paul campus of the University of Minnesota, a visitor center at the Minnesota Historical Society in Fort Snelling, the Walker Library in Minneapolis, the visitors center at the United States Air Force Academy in Colorado Springs, and a

300-acre light manufacturing, office, and warehouse project in Mount Prospect, Illinois.

Among the earth-sheltered projects yet to be translated from drawing board to construction site are the following:

- An earth-sheltered annex for the state capitol building in St. Paul, Minnesota has been selected in a national competition. This $36-million structure would occupy 380,000 square feet and be built beneath the mall in front of the present state capitol. Designed to preserve the landscape as well as save energy, the project awaits adequate funding.
- New York architect Ada Karmi-Melamede has proposed a Second Avenue Spine Project for New York. This massive underground complex would attempt to relieve surface congestion on Manhattan's East Side by combining both residential and commercial space in an attractive, sunlit environment.
- The city of St. Paul, Minnesota is thinking of permitting an abandoned stretch of freeway right-of-way to be converted into an earth-sheltered housing project. If built, a five-mile Mississippi River bluff would be dotted with 1,770 earth-sheltered dwellings in varying densities.

The idea of building earth-sheltered tract homes is not new. In the mid-1960s, architect Richard Kaplan proposed a fifty-six-acre cluster of earth-sheltered homes near Southampton, Long Island. Despite the fact that the fifty-two-home project featured attractive landscaping and plenty of open space, it was never built. By mid-1980, mass-produced earth-sheltered homes remained a concept rather than a reality, although several projects have been proposed.

Barring a prolonged slump in the housing industry, the 1980s will surely herald the advent of the mass-produced earth-sheltered home. What will underground tract homes be like? Three theoreticians of the earth-shelter movement have spent considerable time pondering this question, suggesting various ways to best integrate such housing with the environment. Seen through their eyes, mass-produced earth-sheltered homes offer a bright promise not only in terms of energy savings, but also as a way of improving the overall quality of life.

FRANK MORELAND

One of the most prominent national proponents of earth-sheltered architecture, Frank Moreland received his master's degree in architecture at the University of California at Berkeley in 1967. Since then he has chaired a number of earth-shelter conferences and been associate professor at the school of architecture and environmental design at the University of Texas at Arlington. Today, he heads his own firm, Moreland Associates in Fort Worth, Texas, which deals primarily in earth-sheltered designs.

According to Moreland, energy and transportation constraints will mean a greater emphasis upon medium-density housing in the decades to come. For many Americans, the prospect of living in medium-density housing is not particularly appealing. Open space and green areas tend to be small or nonexistent, yard size is reduced, and visual and acoustical privacy limited.

To overcome these objections, Moreland proposes earth-sheltered urban communities called "hill villages." Complete minicommunities, these hill villages would contain their own post office, stores, parks, and service areas. Vehicular traffic would be banned, and Moreland claims the increased pedestrian traffic would foster a sense of community. Hill villages would also be of sufficient density to support viable mass transit, thus saving energy by providing an alternative to the automobile.

Within such an earth-sheltered community, Moreland envisions a variety of structures, including rental units, single-family dwellings, and underground schools, libraries, and athletic facilities. Property rights could be altered to reflect the novel environment. An individual might own his or her home, but the roof might be retained under neighborhood control for park or recreational use. According to Moreland, the advantages of the hill village concept are numerous:

- Heating and cooling needs reduced by two-thirds
- Greater security
- Reduction in maintenance costs
- Reduced construction time
- More privacy and less noise
- Less long-term environmental damage

By any definition, Frank Moreland is an earth-shelter optimist. He predicts that by 1985 home owners will be able to purchase earth-sheltered tract homes for two-thirds the cost of a tract home in 1980. He also predicts that earth-covered settlements could go a long way toward revitalizing urban areas, rectifying many of the problems which plague the nation's cities. Concludes Moreland, "These are buildings our grandchildren would wish us to build."

THOMAS BLIGH

While teaching at the University of Minnesota, Dr. Thomas Bligh became a well-known champion of the earth-shelter cause. Still a leading underground spokesman, Bligh is now an associate professor of mechanical engineering at the Massachusetts Institute of Technology.

Like Moreland, Bligh has spent many hours contemplating the prospects of a residential earth-sheltered community. He envisions an entire

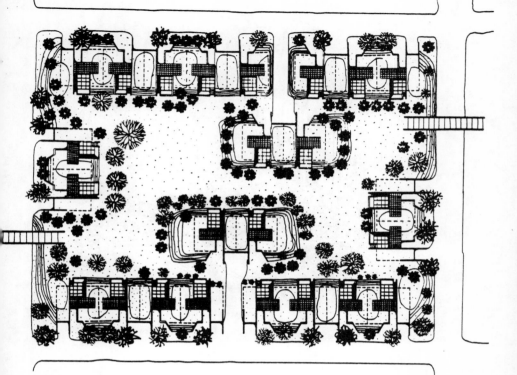

Thomas Bligh's plan for a city block of earth-sheltered homes. (Courtesy: Thomas Bligh.)

city block comprised of three-bedroom earth-sheltered homes. The sod-roofed living room would be above grade, with the bulk of the structure buried underground. Each house would have its own courtyard and face a central park. These green areas would be linked by bridges spanning streets and highways.

Such homes could be constructed comparatively quickly and inexpensively, insists Bligh. Large construction machinery would excavate huge trenches, and the prefabricated homes would be assembled on site. After assembly, the earth would be pushed back and then landscaped. The result would be a mass-produced housing project which would avoid the visual repetition so common in suburban-style homes. Unlike most contemporary tract projects, underground homes would also enjoy a quiet, near-rural tranquility despite their convenient urban location.

Perhaps the most attractive feature of Bligh's plan is that the density would not be achieved at the expense of open land. Bligh estimates that a high-density diagonally developed earth-sheltered community would be 21 percent denser than the most crowded areas of St. Paul, Minnesota, but with 250 percent more green space. Such a community could reduce its overall energy needs by a staggering 84 percent.

ENTRANCE

STUDY

LIVING ROOM

KITCHEN

BATH

BEDROOM

This earth-sheltered home was the first of many designed by architect Malcolm Wells. (Courtesy: Malcolm Wells, from Underground Designs.)

MALCOLM WELLS

Malcolm Wells has persisted in expounding his architectural philosophy. His book *Underground Designs* contains dozens of sketches of earth-sheltered homes, many of them soaring flights of fancy which, although never built, served to inspire others to investigate the promise of earth-sheltered architecture.

His role as a prophet firmly secure, Wells sees a bright future for mass-produced earth-sheltered homes. He predicts "inhabited hills" will become commonplace—terraced homes tucked into hillsides, covered with natural vegetation. Some of these communities could be built on the lower slopes of mesas. Energy would be generated from the sun and wind, and food grown on nearby flood plains. Most vehicular traffic would be banned. Wells also envisions urban earth-sheltered communities featuring solar-heated homes priced for the low- to middle-income buyer. Abundant park land would replace the monotony of asphalt and concrete so common in residential communities today.

As Wells wrote in the February 1976 issue of *The Futurist*, ". . . underground architecture offers us buildings truly in harmony with our planet, buildings that improve with age the way a sapling does, and change with all the seasons: beautiful, living architecture-of-the-earth."

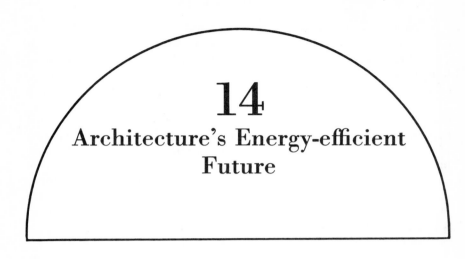

14
Architecture's Energy-efficient Future

Alfred Mercier once wrote, "There was a wise man in the East whose constant prayer was that he might see today with the eyes of tomorrow."

Unfortunately, none of us is blessed with such insight into the future. As unfathomable as a sorcerer's riddle, the future yields its secrets only when once-distant tomorrows become discernible todays. Prognosticators engage in a perilous pastime, and predicting the course of an architectural movement is a risky venture indeed. The risks are greater still in so new a movement as earth-sheltered architecture.

This book would not be complete without an assessment of the direction in which the earth-shelter movement is headed. What kinds of buildings will we construct in the future? How responsive will they be to the preservation of dwindling natural resources? What will be the role of earth-sheltered buildings in the years to come? Will the earth-shelter option blossom or be dismissed as a practical but faddish pursuit?

If the United States is to achieve its stated goal of energy independence, it will need to conduct an all-out campaign to conserve energy. The development of synthetic fuels, a greater reliance on coal and nuclear energy, and stepped-up efforts to discover new oil in the United States are all time-consuming, costly, and environmentally risky. Conservation is both practical and risk-free. Conservation means an immediate reduction in energy consumption. It means saving money, too, since it is cheaper to conserve energy than to produce it. Conservation neither harms nor alters the environment. The implications of a full-scale national conservation effort are far-reaching, particularly in the building sector.

In 1975–1976, the American Institute of Architects Research Corporation studied the energy efficiency of various American buildings constructed during those years. Their conclusion: with existing technology, such buildings could have saved 30 to 50 percent of the energy they consumed. An estimated 5 to 7 million barrels of oil could be saved each day by 1990 if existing buildings took simple steps to increase their energy efficiency. According to a study by the National Bureau of Standards, increasing the thermal performance of residential buildings in the United States could save $100 billion in energy costs by the year 2000.

An architect or builder needn't bury a building underground to realize energy savings. Improved insulation, proper site planning, commonsense landscaping, and the increased use of devices such as storm windows and awnings go a long way toward reducing residential energy consumption. But the earth-shelter option does yield easier, more substantial savings in energy than building on the surface. Earth-sheltered buildings work with nature, not against it. The little extra energy required during the construction of an earth-sheltered building is offset by the appreciable savings over the structure's life span.

Malcolm Wells predicts that by the end of the century as many as 10 million earth-sheltered buildings will be buried from coast to coast. According to Wells, the vast majority will be underground homes. Should such an optimistic projection prove correct, the earth-shelter alternative would make a significant dent in overall building energy consumption in the United States. Yet if Wells's prediction is to become a reality, the earth-shelter movement will have to overcome professional, governmental, legal, and psychological obstacles.

Consider the role of architects, for example. Although only 15 percent of single-family homes in the United States is designed by architects, design professionals carry considerable clout. An architect-designed home garners more attention than one constructed by a builder. Such homes also influence the nation's builders, who look to the architectural profession for inspiration and direction.

But according to Malcolm Wells, many contemporary architects fear earth-sheltered buildings, equating them with "nonarchitecture." Others have yet to be convinced of the seriousness of the earth-shelter option. "By and large," says Michael Barker, former administrator of AIA's Department of Practice and Design, "most architects look at earth-sheltered buildings as something of a curiosity." Many of the architects who have become earth-shelter converts have embraced the concept enthusiastically, but resistance within the profession remains widespread. Despite its rich historical legacy, earth sheltering is perceived as too radical an option by many mainline architects. Tradition dies hard in any profession, especially one whose roots have been on top of the earth rather than beneath it.

For many architects, a building is a "statement." The classic American home is large, impressive, lavish—a residence which can be readily seen and envied by neighbors and passers-by. This adherence to image also exists in commercial architecture. Large business firms are notorious for erecting brash high-rises to function as corporate symbols; a good example is the pyramid-shaped Transamerica tower in San Francisco.

If such buildings are statements, then it is time that architects and clients to consider what kinds of statements they are. All too often, the message is abundantly clear: "I'm a bold, flashy energy guzzler with little respect for either the environment or the people who occupy me."

The earth-sheltered building speaks a different voice. By making an architectural nonstatement, it symbolizes an integration and harmony with its environment. It shows a respect for dwindling energy supplies and a concern for energy conservation. The underground building whispers a subdued but relevant message, one which says as much about the architect and owner as the building itself.

Just as the public needs to be educated to the advantages of earth-sheltering, so too do architects and other professionals. The technical obstacles to underground design and construction are not great. Although the technology is proven and readily available, there are a number of technical considerations which are unique to earth-sheltered construction. According to Dr. Lester Boyer, professor of architecture at Oklahoma State University, a growing number of architects, engineers, and builders are interested in earth-sheltered buildings, but fear they lack the necessary expertise. Without a wider dissemination of earth-shelter technical data, there is a very real possibility that public pressure for earth-sheltered homes may overwhelm professionals' ability to provide them.

A variety of governmental and jurisdictional hurdles need to be overcome also. On the local level, zoning ordinances are often overly restrictive. According to Ray Sterling, director of the Underground Space Center, some zoning codes essentially prevent the construction of earth-sheltered buildings. As architect Kenneth Labs has aptly observed, local zoning ordinances often promote uniformity and discourage novel building designs.

Zoning ordinances have a purpose; manufacturing plants don't belong in residential communities. By controlling land use and regulating density, zoning helps protect the public. But this protection should not be at the expense of esthetically appealing, energy-efficient buildings. Too often, earth-sheltered homes are labeled "basements" rather than recognized as bona-fide buildings. Minnesota has passed laws stating that no local zoning ordinance can indiscriminately zone out earth-sheltered construction. Similar legislation is needed elsewhere in the nation.

Likewise, building codes need to be modified to speed the development of earth-sheltered construction. The intent of building codes is laudable; safe, well-designed buildings are unquestionably in the public interest. Yet as James Scalise notes in *Earth Integrated Architecture*, "Many controls are outdated in purpose, minimal in intent, and unresponsive to change." Building codes need to reflect the realities of an energy-short America, and part of that recognition involves consideration of unconventional designs, earth-sheltered among them.

On a broader scale, the earth-shelter movement could benefit greatly by increased support from elected officials. In Minnesota, state legislators have generally been very supportive. They recognize that the state can trim energy use by tapping a vital and little-used resource: underground space. Support has gone beyond lip service; Minnesota legislators have provided the dollars necessary to fund the Underground Space Center, the Minnesota Housing Finance Agency demonstration project, and much-needed local research in the field. The long-term payback in energy savings and environmental impact may well prove more substantial than the original investment.

To date, the federal role in assisting the earth-shelter movement has been minimal. The Departments of Energy, and Housing and Urban Development are aware of the energy savings which underground structures provide, and both have funded small-scale research projects in the field. However, neither has allocated sufficient sums to provide long-term, in-depth studies which could promote a wider proliferation of earth-sheltered structures. Nor has Congress recognized the energy-saving potential of such buildings by enacting tax-credit legislation as an incentive to further earth-sheltered construction.

Legislation is also needed to reflect the country's growing reliance on solar energy. In ancient Rome, it was illegal to erect any building which blocked a neighbor's sunlight; as a result, Roman home owners were assured of an uninterrupted supply of passive solar energy. Few such laws exist in the United States today. Since so many earth-sheltered homes rely on either passive or active solar energy to trim fuel consumption, the question of "solar rights" is one which will need to be addressed in the years ahead.

Perhaps the biggest obstacle which faces the earth-shelter movement is the psychological one, the bias discussed in Chapter 2. Certainly earth-shelter consciousness is on the rise. The subject is being discussed in the media, and more and more Americans are reevaluating their negative feelings toward underground space. In addition, an increasing number of earth-sheltered homes are being made available for public inspection. "This public visibility is very important," insists Mary Tingerthal, coordi-

In urban areas, building codes and zoning ordinances sometimes prevent construction of earth-sheltered homes. Above, William Morgan's Forest House in central Florida. (Courtesy: William Morgan Architects.)

nator of the special housing program for the Minnesota Housing Finance Agency. "We have to get houses out there that people can look at, tour, and feel for themselves."

Often the strongest resistance to earth-sheltering comes from the very people who could most influence such construction: lenders. Their support is crucial if earth-sheltered homes are to proliferate. "The major decisions on housing in this country are made by the lending institutions, not designers or builders or even the occupants," argues John Selfridge, associate professor of architecture and design at Kansas State University. Some lenders are willing to consider more novel designs; lenders in Minneapolis, Fort Worth, and parts of Oklahoma are familiar with the advantages of such homes and generally cooperate in helping an owner obtain a mortgage. More conservative lenders scoff at earth-sheltered designs, dismissing them as a fad and a poor financial risk.

According to Tingerthal, one of the lenders' major concerns is determining how to obtain an appraisal on such a property. Because earth-sheltered homes are a recent phenomenon, few comparisons exist. Earth-sheltered homes have yet to establish a discernible resale record in the marketplace. Unwilling to provide a mortgage for what they consider a risky design, many lenders deny the request outright.

Earth-shelter mortgages would be easier to obtain if lenders took into account the demonstrable energy savings which result. If an automobile with a mileage of seventy-five miles per gallon were introduced, no one would dispute that it is an excellent investment, one with a high resale potential. Similarly, earth-sheltered homes can save their owners thousands of dollars on utility costs over the life-cycle of the building, because they consume so little energy.

"A lot depends on how enlightened the lender is," says Tingerthal. "Not everyone believes there's an energy crisis. Some won't know we've run out of oil until the last drop runs dry." For this reason, prospective home owners often must come armed with exceptionally good plans, private appraisals, and a stronger-than-usual equity position to convince lenders that earth-sheltered homes are as mortgage-worthy as an energy gulper on the surface.

Assuming these obstacles can be overcome, there seems no reason to doubt that the growing popularity of earth-sheltered dwellings can be sustained in the years to come. A wide variety of underground homes, offices, schools, churches, libraries, and museums already dot the American landscape. In the future, hospitals, factories, sports arenas, and utility plants will benefit from the same subterranean location. Frank Moreland even suggests a bold earth-shelter plan to restructure the nation's cities: a massive federal commitment comparable to the space program which would create underground urban communities as a viable alternative to the wasteful, sorry housing which plagues American cities today.

American architecture seems ripe for another revolution. This upheaval will be spurred not by architects, but by clients. Unwilling to pay spiraling utility costs and unhappy with existing designs, Americans will demand more energy-efficient and environmentally "gentle" buildings. As a result, the regional design ethic will reappear, and architects will either master the intricacies of energy conservation or risk losing clients.

Earth-sheltered buildings will be an integral part of this new direction in architecture. Not every structure will be buried underground; nor should it be. The earth-sheltered design is not an architectural cure-all for the energy crisis. It is an option, not a panacea. But a greater proliferation of earth-sheltered buildings will trigger a new awareness of man's relationship with the environment. Such buildings will be seen not as so much wood and concrete, but extensions of the earth around them. When such an awareness prevails, people will no longer ask why should we build earth-sheltered buildings, but why not?

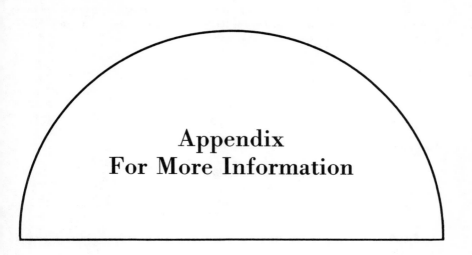

Appendix
For More Information

UNIVERSITY CLEARING-HOUSES

Six major American universities provide a variety of information on earth-sheltered construction for both the lay public and architects, engineers, builders, and other professionals.

The following three schools offer printed materials and earth-shelter seminars. They are able to respond to specific technical questions. Enclose a self-addressed stamped envelope when requesting leaflets or brochures.

Underground Space Center
11 Mines and Metallurgy
221 Church Street, S.E.
University of Minnesota
Minneapolis, MN 55455

Architectural Extension
115 Architecture Building
Oklahoma State University
Stillwater, OK 74078

Clearinghouse for Earth Covered Settlements
School of Architecture and Environmental Design
University of Texas at Arlington
Arlington, TX 76019

Although smaller in size, the following three schools can respond to technical queries and provide information on local earth-shelter architects and builders. Write:

Dr. E. W. Kiesling
Civil Engineering Department
Box 4089
Texas Tech University
Lubbock TX 79409

(Also offers one-day earth-shelter seminars for the public.)

Dr. Truman Stauffer
Geosciences Department
University of Missouri
Kansas City, MO 64110

(Dr. Stauffer is an expert on the commercial use of limestone caverns beneath Kansas City.)

Dr. Nolan Aughenbaugh
University of Missouri at Rolla
Rolla, MO 65401

SOLAR INFORMATION

For detailed information on solar energy as well as a variety of books and other solar publications, write:

National Solar Heating and Cooling
Information Center
P.O. Box 1607
Rockville, MD 20850
Phone toll-free: (800) 523-2929
In Pennsylvania: (800) 462-4983

PERIODICALS

Underground Space, a professional journal dedicated to "encouraging analysis, research, and distribution of information on technical, environmental, legal, economic, social, and political problems associated with the use of the underground."

Available at an annual rate of $62; two-year rate, $117.80.

The American Underground Space Association
Department of Civil and Mineral Engineering
University of Minnesota
Minneapolis, MN 55455

Earth Shelter Digest & Energy Report, an illustrated bimonthly publication containing features and new developments in the field of earth-sheltered buildings. Articles discuss design, construction, ventilation, waterproofing, financing, orientation, solar applications, and similar subjects. Also includes a number of case studies of earth-sheltered homes throughout the country. A must for the serious earth-shelter enthusiast. Subscription cost: $15 annually.

> Earth Shelter Digest & Energy Report
> Webco Publishing, Inc.
> 1701 East Cope
> St. Paul, MN 55109

BOOKS

The Architectural Use of Underground Space: Issues and Applications. Kenneth Labs. 1975. Kenneth Labs, Star Route, Mechanicsville, PA 18934. Cost: $20.

Architecture Underground. Kenneth Labs. Architectural Record Books. Available late summer, 1981.

Building Underground for People: Eleven Selected Projects in the United States. Michael Barker. 1978. American Institute of Architects, 1735 New York Avenue, N.W., Washington, DC 20006. Cost: $6.

Earth Covered Buildings. Frank Moreland, editor. Proceedings of Conference held in Fort Worth, Texas, July 9–12, 1975. Superintendent of Documents, U.S. Government Printing Office, Washington, DC 20402. Stock No. 038-000-00286-4. Cost: $3.25.

Earth Integrated Architecture. James W. Scalise, editor. 1975. Architecture Foundation, College of Architecture, Arizona State University, Tempe, AZ 85281. Cost: $10.

Earth Shelter Housing Design. Prepared by Underground Space Center, University of Minnesota. 1978. Van Nostrand Reinhold Company. Underground Space Center, 11 Mines and Metallurgy, 221 Church Street, S.E., University of Minnesota, Minneapolis, MN 55455. Cost: $11.

Gentle Architecture. Malcolm Wells. 1980. McGraw-Hill.

Notes from the Energy Underground. Malcolm Wells. 1980. Van Nostrand Reinhold.

Underground Designs. Malcolm Wells. 1977. Malcolm Wells, P.O. Box 1149, Brewster, MA 02631. Cost: $6.

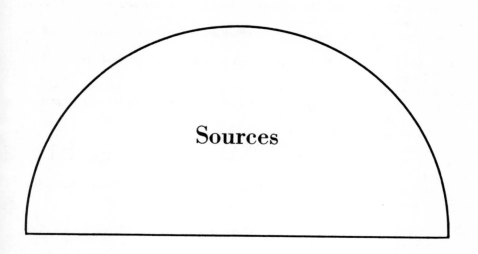

Sources

CHAPTER 1

"CAVE." *Collier's Encyclopedia* (1979), vol. 5, pp. 586–89.

HAZER, FAHRIYE. "Cultural-Ecological Interpretation of the Historical Underground Cities of Goreme, Turkey," in *Alternatives in Energy Conservation: The Use of Earth Covered Settlements*. Washington, D.C.: U.S. Government Printing Office, 1975, pp. 21–28.

"HISTORY IS WITH US." *Progressive Architecture*, vol. 48, April 1967, pp. 138–51.

KELLER, CHARLES. "Cave Dwellers." *Encyclopedia Americana* (1979), vol. 6, pp. 105–106.

LABS, KENNETH. "The Architectural Underground," *Underground Space*, vol. 1, 1976, pp. 2–4.

LANIER, ROYCE. *Geotecture*. University of Notre Dame, 1971, pp. 3–6, 10, 12, 28–30, 32–34.

MOHR, CHARLES. "Cave." *Encyclopedia Americana* (1979), vol. 6, pp. 101–105.

RUDOFSKY, BERNARD. *The Prodigious Builders*. New York: Harcourt Brace Jovanovich, 1977, pp. 21–47.

CHAPTER 2

"THE CAVEMEN." *Time*, 23 June 1958, p. 22.

COLLINS, BELINDA. *Windows and People: A Literature Survey*. Washington D.C.: National Bureau of Standards, June 1975.

DEAN, ANDREA. "Underground Architecture." *American Institute of Architects Journal*, vol. 67 no. 4, April 1978.

DEMPEWOLFF, RICHARD. "Underground Housing." *Science Digest,* vol. 78 no. 5, November 1975, pp. 40–53.

FAIRHURST, CHARLES. "Going Under to Stay on Top." *Underground Space,* vol. 1, 1976, pp. 71–86.

LABS, KENNETH. "The Architectural Underground." *Underground Space,* vol. 1, 1976, pp. 5–8.

LANIER, ROYCE. *Geotecture.* University of Notre Dame, 1971, pp. 26–27, 47–50.

PAULUS, PAUL. "On the Psychology of Earth Covered Buildings." *Underground Space,* vol. 1, 1976, pp. 127–30.

"ROOM AT THE TOP." *Newsweek,* 3 May 1965, p. 87.

"THE SHELTERED LIFE." *Time,* 20 October 1961, pp. 21–25.

SUGGS, ROBERT. "Bomb Shelter." *Encyclopedia Americana* (1979), vol. 4, pp. 186–87.

"WORLD'S FAIR VISITORS VIEW LIFE UNDERGROUND." *Today's Health,* September 1964, p. 12.

CHAPTER 3

DEAN, ANDREA. "The Underground Movement Widens." *American Institute of Architects Journal,* vol. 67 no. 13, November 1978, pp. 34–50.

"DESIGN BRINGS ARCHITECT SPECIAL ATTENTION." *Earth Shelter Digest & Energy Report,* vol. 4, July/August 1979, pp. 44–45.

"EAST COAST'S EARLIEST IS MOST COPIED." *Earth Shelter Digest & Energy Report,* vol. 7, January/February 1980, pp. 37–39.

Ecology House. Pamphlet distributed by John Barnard.

EDELHART, MIKE. "The Good Life Underground." *Omni,* January 1980, pp. 50–55.

Interview with JOHN BARNARD, September 1978.

Interview with DON METZ, September 1978.

Interview with MICHAEL MCGUIRE, September 1978.

Interview with MALCOLM WELLS, March 1980.

"JOHNSON UNDERGROUND." *Progressive Architecture,* vol. 48, April 1967, pp. 138–51.

LANIER, ROYCE. *Geotecture.* University of Notre Dame, 1971, p. 37.

Letter to author from Philip Johnson, 14 September 1979.

SANOFF, HENRY. "Seven Acres of Underground Shelter." *American Institute of Architects Journal,* vol. 47 no. 2, February 1967, pp. 66–68.

"SAVING BY GOING UNDERGROUND." *American Institute of Architects Journal,* vol. 61 no. 2, February 1974, pp. 48–49.

SCALISE, JAMES, ed. *Earth Integrated Architecture.* Arizona State University, 1975, pp. G-11, G-12, G-17.

SCULLY, VINCENT. *American Architecture and Urbanism.* New York: Praeger, 1969.

SMAY, ELAINE. "Underground Living." *Popular Science,* vol. 204, June 1974, pp. 88–89.

UNDERGROUND SPACE CENTER. *Earth Sheltered Housing Design.* New York: Van Nostrand Reinhold, 1978, pp. 201, 208, 215, 248.

"UNIQUE TUBE HOME FITS INTO THE EARTH." *Earth Shelter Digest & Energy Report,* vol. 1, January/February 1979, pp. 20–22.

WELLS, MALCOLM. "The Absolutely Constant Incontestably Stable Architectural Value Scale." *Progressive Architecture*, vol. 52 no. 3, March 1971, pp. 92–97.

WELLS, MALCOLM. "Environmental Impact." *Progressive Architecture*, vol. 55 no. 6, June 1974, pp. 59–63.

WELLS, MALCOLM. "Nowhere to Go but Down." *Progressive Architecture*, vol. 46, February 1965, pp. 174–79.

WELLS, MALCOLM. "To Build without Destroying the Earth," in *Alternatives in Energy Conservation: The Use of Earth Covered Settlements*. Washington, D.C.: U.S. Government Printing Office, 1975, pp. 211–32.

WELLS, MALCOLM. "Underground Architecture." *CoEvolution Quarterly*, Fall 1976, pp. 85–94.

WELLS, MALCOLM. "Why I Went Underground." *The Futurist*, February 1976, pp. 21–24.

"WINSTON HOUSE, Lyme, New Hampshire." *Architectural Record*, vol. 155, Mid-May 1974, pp. 52–63.

CHAPTER 4

"BEFORE THE VIRGIN MET THE DYNAMO." *Architectural Forum*, July/August 1973, pp. 77–85.

"BEHIND THE REVOLT AGAINST MODERN ARCHITECTURE." *U.S. News & World Report*, 3 September 1979, p. 62.

BROLIN, BRENT. *The Failure of Modern Architecture*. New York: Van Nostrand Reinhold, 1976.

"BUILDINGS OF FUTURE WILL SAVE FUEL." *U.S. News & World Report*, 18 February 1974, pp. 47–48.

"BUILDINGS THAT DON'T SQUANDER ENERGY." *Popular Science*, June 1974, pp. 84–87.

"DOING THEIR OWN THING." *Time*, 8 January 1979, pp. 52–59.

"THE ENERGY ISSUE." *Progressive Architecture*, vol. 60, April 1979, p. 8.

GOLDBERGER, PAUL. "Architectural Soup." *New York Times*, 30 December 1979, p. D41.

"IS THE ENERGY CRISIS FOR REAL?" *Architecture Minnesota*, March/April 1979, pp. 18–26.

LYNCH, MITCHELL. "Moves to Save Energy Appear to be Faltering in Architectural Field." *Wall Street Journal*, 26 September 1977, p. 1.

STEIN, RICHARD. *Architecture and Energy*. New York: Anchor Books, 1978, pp. 28–47.

UPJOHN, EVERARD. "Architecture." *Encyclopedia Americana* (1979), vol. 2, pp. 217–38.

WATSON, DONALD, ed. *Energy Conservation Through Building Design*. New York: McGraw-Hill, 1979, pp. 1–3, 10, 23, 26, 27, 37, 38, 53, 97, 98.

CHAPTER 5

BLIGH, THOMAS. "A Comparison of Energy Consumption in Energy Covered vs. Non–earth Covered Buildings." *Alternatives in Energy Conservation: The Use of Earth Covered Settlements*. Washington, D.C.: U.S. Government Printing Office, 1975, pp. 85–104.

BLIGH, THOMAS and HAMBURGER, RICHARD. "Conservation of Energy by Use of Underground Space," in *Legal, Economic and Energy Considerations in the Use of Underground Space*. National Academy of Sciences, 1974, pp. 103–17.

Clearinghouse for Earth Covered Settlements, University of Texas at Arlington. *Fact Sheet on Earth Sheltered Construction.*

DEMPEWOLFF, RICHARD. "Your Next Home Could Have a Grass Roof." *Popular Mechanics,* March 1977, pp. 78–81, 142–46.

HAUPERT, DAVID. "Underground Housing Is Coming on Strong." *Better Homes & Gardens,* September 1979, pp. 97–102, 180.

LANIER, ROYCE. *Geotecture.* University of Notre Dame, 1971, pp. 42–44, 51.

LYTTON, ROBERT. "Soil and Water Considerations," in *Alternatives in Energy Conservation: The Use of Earth Covered Settlements.* Washington, D.C.: U.S. Government Printing Office, 1975, pp. 257–62.

"REDUCE THE LOAD: The Building as a Heat Exchanger." *Progressive Architecture,* April 1979, pp. 80–82.

REINDL, WILHELM. "Getting into It." *Earth Shelter Digest & Energy Report,* vol. 7, January/February1980, pp. 16–17.

SIMMONS, LON. "Success with Residential Post-tensioning." *Earth Shelter Digest & Energy Report,* vol. 4, July/August 1979, pp. 23–27.

SMAY, ELAINE. "Underground Houses." *Popular Science,* April 1977, pp. 84–89.

STERLING, RAY. "Structural Systems for Earth Sheltered Housing." *Underground Space,* vol. 3 no. 2, 1978, pp. 75–81.

UNDERGROUND SPACE CENTER. *Earth Sheltered Housing Design.* New York: Van Nostrand Reinhold, 1978.

WOLF, RAY. "The Good Feeling of Living in the Earth." *Organic Gardening,* December 1978, pp. 58–65.

CHAPTER 6

"ARCHITECT DETAILS WITH WOOD." *Earth Shelter Digest & Energy Report,* vol. 5, September/October 1979, pp. 17–19.

BALCOMB, J. DOUGLAS. "The Heat also Rises." *Progressive Architecture,* vol. 60, April 1979, pp. 106–109.

"CAN SUN POWER PAY?" *Newsweek,* 18 December 1978, pp. 66–70.

CHASE, VICTOR. "Active or Passive?" *Science Digest,* April 1979, pp. 62–66.

"DO-IT-YOURSELFERS GIVE SOLAR ENERGY A SHARP BOOST." *U.S. News & World Report,* 17 October 1977, pp. 96–98.

GREENE, WADE. "Solar Refractions." *New Times,* 29 May 1978, pp. 4–6.

Interview with DON METZ, September 1978.

LOFTNESS, VIVIAN and REEDER, BELINDA. "Recent Work in Passive Solar Design." *American Institute of Architects Journal,* vol. 67, April 1978, pp. 52–63.

ODDO, SANDRA. "Solar: Trial and Error." *Progressive Architecture,* vol. 60, April 1979, pp. 98–102.

ROGERS, PATSY. "Underground." *Washington Star Sunday Magazine,* 1 January 1978, pp. 10–12.

SHICK, WAYNE. "Proper Building Orientation Can Save You Energy." *Earth Shelter Digest & Energy Report,* vol. 2, March/April 1979, pp. 38–39.

SOLAR ENERGY AND YOUR HOME. U.S. Department of Housing and Urban Development, August 1979.

"SOLAR ENERGY HEADS FOR THE BIG LEAGUES." *U.S. News & World Report*, 8 May 1978, pp. 58–59.

"THE SOLAR REVOLUTION." *Newsweek*, 7 April 1980, pp. 79–85.

"THE SUN STARTS TO RISE ON SOLAR." *Time*, 1 May 1978, pp. 67–68.

WATSON, DONALD. *Energy Conservation Through Building Design.* New York: McGraw-Hill, 1979, p. 8.

WEBSTER, BAYARD. "Research Finds Greeks Used Solar Energy." *New York Times*, 24 April 1979, p. C2.

"WHAT'S AHEAD FOR SOLAR ENERGY?" *U.S. News & World Report*, 3 March 1980, pp. 53–54.

WILHELM, JOHN. "Solar Energy, the Ultimate Powerhouse." *National Geographic*, March 1976, pp. 381–97.

CHAPTER 7

BARKER, MICHAEL. *Building Underground for People.* American Institute of Architects, 1978.

BENNETT, DAVID. "University of Minnesota Book Store," in *Alternatives in Energy Conservation: The Use of Earth Covered Settlements.* Washington, D.C.: U.S. Government Printing Office, 1975, pp. 117–30.

BENNETT, DAVID and BLIGH, THOMAS. "The Energy Factor—A Dimension of Design." *Underground Space*, vol. 1, 1977, pp. 325–32.

CARTER, DOUGLAS. "Terraset School." *Underground Space*, vol. 1, 1977, pp. 317–28.

Change and Continuity in the Harvard Yard: The Nathan Marsh Pusey Library. Harvard University, 1976.

DEAN, ANDREA. "Evaluation: Bright Bookstore Cut into the Earth." *American Institute of Architects Journal*, vol. 67 no. 4, April 1978, pp. 46–50.

DEAN, ANDREA. "Two Communities Build Subsurface Schools." *American Institute of Architects Journal*, vol. 67 no. 13, November 1978, pp. 47–49.

The Effect of Windowless Classrooms on Elementary School Children. Architectural Research Laboratory, Department of Architecture, University of Michigan, 1965.

FRAM, MARCIA. "Saving Energy at a Reston School." *Washington Star*, 1 March 1980, p. B1.

"GEORGETOWN UNIVERSITY COVERS FIELD HOUSE WITH FIELD." *Earth Shelter Digest & Energy Report*, vol. 8, March/April 1980, pp. 44–46.

"GOING UNDERGROUND." *Progressive Architecture*, vol. 48, April 1967, pp. 138–51.

"HIDING OUT IN HARVARD YARD." *Interior Design*, vol. 47, December 1976, p. 144.

"IN PROGRESS." *Progressive Architecture*, vol. 60, April 1979, p. 40.

Interview with THOMAS BLIGH, September 1978.

Interview with PAUL SCHIPP, Williamson Hall, September 1978.

Interview with TONY MARTIN, Terraset School, September 1978.

Letter to author from William Scott, Director, University of Houston Center, 25 September 1979.

LUTZ, FRANK. *Abo Revisited.* Defense Civil Preparedness Agency, 1972.

LUTZ, FRANK. "Studies of Children in an Underground School." *Underground Space*, vol. 1, 1976, pp. 131–34.

"NEW DIGS." *Time,* 5 February 1965, p. 47.

"OKLAHOMA MAKES CASE FOR UNDERGROUND SCHOOLS." *Earth Shelter Digest & Energy Report,* vol. 8, March/April 1980, pp. 40–42.

"THE STUDENT UNDERGROUND." *Progressive Architecture,* vol. 52, April 1971, pp. 78–85.

Terraset. Pamphlet prepared by Terraset Elementary School, 1977.

University of Illinois Undergraduate Library. Fact sheet prepared by University of Illinois at Champaign-Urbana.

CHAPTER 8

"BANKERS BUILD SECURELY INTO ENVIRONMENT." *Earth Shelter Digest & Energy Report,* vol. 3, May/June 1979, p. 37.

"BELOW-GROUND PLAN LOWERS LIBRARY COSTS." *Building Design & Construction,* June 1979, pp. 14–15.

"BIGGEST BIG TOP." *Architectural Forum,* December 1969, pp. 68–69.

"BOLD DESIGN PAYS OFF ON ENERGY." *American School & University,* December 1976, pp. 22–25.

DEAN, ANDREA. "Church Bermed to Diminish its Impact." *American Institute of Architects Journal,* vol. 67 no. 4, April 1978, pp. 42–43.

"ESPIRIT GROWS IN BROOKLYN." *Progressive Architecture,* vol. 59, May 1978, pp. 62–65.

Fact Sheets on Underground Annex. Mutual of Omaha.

"FLORIDA MUSEUM USES EARTH FORMS BOLDLY." *Architectural Record,* September 1971, pp. 121–25.

"HIGHWAY REST STOP SAVES LANDSCAPE." *Earth Shelter Digest & Energy Report,* vol. 8, March/April 1980, p. 23.

"HOUSE MONITORING PROJECT TO PROVIDE ENERGY USE DATA." *Underline,* vol. 1 no. 3, March 1980, p. 1.

KOHN, SHERWOOD. "It's OK to Touch at the New-style Hands-on Exhibits." *Smithsonian,* September 1978, pp. 78–83.

Letter to author from Wayland Porter, Minnesota Department of Natural Resources, 24 October 1979.

"MUTUAL OF OMAHA DESIGNS NEW OFFICES." *Earth Shelter Digest & Energy Report,* vol. 5, September/October 1979, pp. 37–39.

SHANK, BEN. "The Earth Shelter Housing Demonstration Project, Minnesota Housing Finance Agency." *Underground Space,* vol. 3 no. 5, 1978, pp. 259–68.

"THE SOLAR UNDERGROUND." *Progressive Architecture,* vol. 60, April 1979, pp. 124–27.

"AN UNDERGROUND CHURCH FOR BENEDICTINE MONKS BY STANLEY TIGERMAN." *Architectural Record,* January 1973, pp. 106–10.

"THE U.S. PAVILION." *Architectural Forum,* April 1970, p. 41.

CHAPTER 9

"ARCHITECTURE AS ENERGY." *Design Quarterly,* February 1977, p. 29.

"BUILDER IN ILLINOIS CUTS FAMILY HEATING COSTS BY MOVING INTO CAVE." *New York Times,* 25 June 1979, p. 18.

DEAN, ANDREA. "Seaside Duplex Molded from a Dune." *American Institute of Architects Journal*, vol. 67 no. 4, April 1978, pp. 38–39.

Earthtec. Fact sheet prepared by Don Metz.

"HE JOINS MOTHER EARTH IN NEW BUILDING VENTURE." *Washington Star*, 25 May 1979, p. F8.

"A HOUSE HEATED SOLELY BY THE SUN." *House & Garden*, October 1979, pp. 176–79.

Interview with CHARLES LANE, Underground Space Center, October 1979.

Interview with MELVIN WILCOX, Davis Caves, October 1979.

Letter to author from Mark Simon, 30 August 1979.

Move Down to Earth Sheltered Townhouse. Pamphlet prepared by Seward West Redesign.

"ON THE FLORIDA COAST." *Architectural Record*, vol. 162 no. 3, Mid-August 1977, pp. 74–75.

RITTER, ALEXANDER. "Demonstration Homes." *Architecture Minnesota*, March/April 1979, pp. 46–47.

"SHELTERED HOME A PROBLEM SOLVER." *St. Paul Dispatch*, 9 April 1979, p. 17.

SLESIN, SUZANNE. "Down to Earth: Two Houses." *New York Times*, 22 March 1979, p. C1.

"TOWNHOUSE FOR SALE." *Earth Shelter Digest & Energy Report*, vol. 5, September/October 1979, p. 12.

UNDERGROUND SPACE CENTER. *Earth Sheltered Housing Design*. New York: Van Nostrand Reinhold, 1978.

"WALLS BUILT INTO THE EARTH HOLD SUN'S HEAT." *Home & Garden Building Guide*, Winter 1978, p. 77.

CHAPTER 10

Interview with TOM and MARY HAGAN, 23 May 1980.

Interview with MARGARET O'CONNOR, 18 May 1980.

Interview with BUDDY WRIGHT, 20 May 1980.

CHAPTER 11

ALTHOFF, SHIRLEY. "The Underground World of Kansas City." *St. Louis Globe-Democrat Sunday Magazine*, 8 May 1977, pp. 8–13.

"BIG CORPORATIONS TAKE OVER." *Progressive Architecture*, vol. 48, April 1967, p. 142.

"CAVEMEN IN KANSAS CITY." *Newsweek*, 1 May 1967, pp. 78–80.

"FINALLY—AN ATTACK-PROOF CENTER FOR U.S. DEFENSE." *U.S. News & World Report*, 24 January 1966, pp. 54–57.

"FORT KNOX FOR RECORDS CARVED OUT OF SALT." *U.S. News & World Report*, 31 July 1978, pp. 52–53.

"INDUSTRY TRIES LIVING IN THE CAVES." *Business Week*, 20 May 1961, pp. 166–70.

Interview with CHERIE WHITE, Great Midwest Corporation, September 1978.

Interview with DR. TRUMAN STAUFFER, University of Missouri, September 1978.

Promotional material supplied by GREAT MIDWEST CORPORATION.

Promotional material supplied by INLAND STORAGE DISTRIBUTION CENTER.

Promotional material supplied by UNDERGROUND VAULTS & STORAGE, INC.

STAUFFER, TRUMAN. *Guidebook to the Occupance and Use of Underground Space in Kansas City.* University of Missouri at Kansas City, 1972.
STAUFFER, TRUMAN. *Underground Space: Inventory and Prospect in Kansas City.* University of Missouri at Kansas City, 1976.

CHAPTER 12

CLAUSEN, OLIVER. "Sweden Goes Underground." *New York Times Magazine,* 22 May 1966, pp. 23–25.
"INVISIBLE ARCHITECTURE IN THE PARIS UNDERGROUND." *Fortune,* January 1966, p. 176.
KEATING, BERN. "The Underground Resurfaces." *American Way,* December 1979, pp. 20–24.
LANIER, ROYCE. *Geotecture.* University of Notre Dame, 1971, pp. 8, 14–16, 26.
"MONTREAL HANDLES WINTER EASILY, COZILY—UNDERGROUND." *Minneapolis Tribune,* 26 December 1976.
MOORE, KENNY. "Coober Pedy: Opal Capital of Australia's Outback." *National Geographic,* October 1976, pp. 560–71.
"PARISIAN THEATER." *Progressive Architecture,* vol. 48, April 1967, p. 148.
RUDOFSKY, BERNARD. *The Prodigious Builders.* New York: Harcourt Brace Jovanovich, 1977, p. 46.
STANFORD, GREG. "Down Under Dugouts—White Cliffs, Australia." *Earth Shelter Digest & Energy Report,* vol. 7, January/February 1980, pp. 41–43.
"SWEDEN DIGS DOWN." *Progressive Architecture,* vol. 48, April 1967, pp. 138–51.

CHAPTER 13

Benham-Blaire Associates, fact sheet on California State Office Building project.
BLIGH, THOMAS. "Energy Conservation by Building Underground." *Underground Space,* vol. 1, 1976, pp. 19–33.
"CONSTRUCTION GOES UNDERGROUND." *The Futurist,* December 1979, pp. 486–88.
ESTES, KIRBY. "The Extra Dimension: Urban Architecture for Tomorrow." *The Futurist,* December 1979, pp. 439–46.
Interview with THOMAS BLIGH, September 1978.
Interview with FRANK MORELAND, September 1978.
Interview with MALCOLM WELLS, March 1980.
JAHN, HELMUT. "Minnesota II Underground Annex to State Capitol Complex." *Underground Space,* vol. 1, 1977, pp. 365–69.
"LARGEST EARTH SHELTER IN SAN FRANCISCO." *Earth Shelter Digest & Energy Report,* vol. 9, May/June 1980, p. 37.
MORELAND, FRANK. "An Alternative to Suburbia," in *Alternatives in Energy Conservation: The Use of Earth Covered Settlements.* Washington, D.C.: U.S. Government Printing Office, 1975, pp. 183–208.
MORELAND, FRANK. "Earth Covered Habitat—an Alternative Future." *Underground Space,* vol. 1, 1977, pp. 295–307.
MORELAND, FRANK. "Notes on Earth Covered Settlements," in *Earth Covered Buildings and Settlements.* National Technical Information Service, 1979, pp. 278–328.
Myers and Bennett Architects/BRW, fact sheet on earth-sheltered projects.

"PRESERVATION OF LANDSCAPE." *Progressive Architecture,* vol. 48, April 1967, pp. 138–51.

SCHURKE, PAUL. "University of Minnesota Builds Down Again." *Earth Shelter Digest & Energy Report,* vol. 6, November/December 1979, pp. 28–29.

TEMKO, ALLEN. "California's New Generation of Energy Efficient State Buildings." *American Institute of Architects Journal,* vol. 66 no. 12, December 1977, pp. 50–56.

CHAPTER 14

"GROWTH OF CENTER REFLECTS INTEREST IN USE OF UNDERGROUND." *Underlines,* vol. 1 no. 1, p. 1.

Interview with DR. LESTER BOYER, 19 May 1980.

Interview with DR. RAY STERLING, 5 May 1980.

Interview with MARY TINGERTHAL, 21 May 1980.

Interview with MALCOLM WELLS, March 1980.

KORELL, MARK. *Financing Earth Sheltered Housing: Issues and Opportunities.* Federal Home Loan Bank Board, 1979.

LABS, KENNETH. "Planning for Underground Housing," in *Underground Utilization: A Reference Manual of Selected Works.* University of Missouri at Kansas City, 1978.

SCALISE, JAMES, ed. *Earth Integrated Architecture.* Arizona State University, 1975, pp. F1–F8.

UNDERGROUND SPACE CENTER. *Earth Sheltered Housing Design.* New York: Van Nostrand Reinhold, 1978, pp. 149–78.

WEBSTER, BAYARD. "Research Finds Greeks Used Solar Energy." *New York Times,* 24 April 1979, p. C2.

WELLS, MALCOLM. "Why I Went Underground." *The Futurist,* February 1976, pp. 21–24.

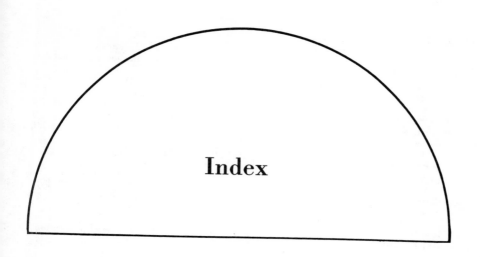

Index

73